ELECTRON MICROSCOPY AND ANALYSIS

THE WYKEHAM SCIENCE SERIES

The aim of the Wykeham Science Series is to introduce the present state of the many fields of study within science to students approaching or starting their careers in University, Polytechnic, or College of Technology. Each book seeks to reinforce the link between school and higher education, and the main author, a distinguished worker or teacher in the field, is assisted by an experienced sixth form schoolmaster.

ELECTRON MICROSCOPY AND ANALYSIS

P. J. Goodhew
Structural Studies Unit
University of Surrey

WYKEHAM PUBLICATIONS (LONDON) LTD
(A MEMBER OF THE TAYLOR & FRANCIS GROUP)
LONDON AND WINCHESTER
1975

First published 1975 by Wykeham Publications (London) Ltd.

Cover illustration—The regular array of faceted grains on the surface of electroplated copper. (G. Gibbs and C. T. Walker.)

ISBN 0 85109 001 X

Printed in Great Britain by Taylor & Francis Ltd.
10–14 Macklin Street, London, WC2B 5NF

Distribution:

UNITED KINGDOM, EUROPE, MIDDLE EAST AND AFRICA

Chapman & Hall Ltd. (a member of Associated Book Publishers Ltd.), North Way, Andover, Hampshire.

WESTERN HEMISPHERE

Springer-Verlag New York Inc., 175 Fifth Avenue, New York, New York 10010.

AUSTRALIA, NEW ZEALAND AND FAR EAST (EXCLUDING JAPAN)

Australia & New Zealand Book Co. Pty. Ltd., P.O. Box 459, Brookvale, N.S.W. 2100.

ALL OTHER TERRITORIES

Taylor & Francis Ltd., 10–14 Macklin Street, London, WC2B 5NF.

PREFACE

PHOTOGRAPHS taken with an electron microscope are increasingly appearing in the pages of introductory science text books and even in the popular press, and many of the other books in the Wykeham Science Series use both transmission and scanning electron micrographs to illustrate structure and morphology on a fine scale. The aim of this volume is to present, at a level which can be understood by those with an 'A' level science background, the principles of the common electron microscope techniques and to illustrate their application to some typical problems.

The text covers the two chief types of electron microscopy (scanning and transmission) and many of their associated techniques such as electron diffraction and microanalysis. Finally a whole chapter is devoted to putting the electron microscope techniques into proper perspective by a discussion of the other possible methods of deducing similar information from a small specimen. It is my hope that the book will be of value both to those studying for science 'A' levels and to those undergraduates in physical, materials and biological sciences who encounter electron microscopy as an essential tool but whose syllabus does not contain a major course on the subject.

It is a pleasure to acknowledge the help I have received from Mrs. Louise Cartwright during the preparation of the manuscript. Without her help I fear that large sections of the text would be unintelligible to the people for whom it is intended. I would also like to thank all those people who have allowed me to use their micrographs and other results. I have tried to acknowledge each individual at an appropriate point, but particular thanks should go to Mike Hepburn and Gill Gibbs who took many of the micrographs specially and Dr. Jim Castle whose comments on Chapter 6 were invaluable. Finally my warmest thanks must go to my mother for typing the whole manuscript both efficiently and quickly and to my wife Gwen for tolerating the loss of so many evenings.

August 1974 PETER GOODHEW

ABBREVIATIONS

AES	Auger electron spectroscopy
CRT	Cathode-ray tube
ECP	Electron channelling pattern
EDS	Energy-dispersive system
EMMA	Electron microscope microanalyser
EPMA	Electron probe microanalyser
ESCA	Electron spectrometer for chemical analysis (=XPS)
FIM	Field ion microscope
HVEM	High voltage electron microscopy
LEED	Low energy electron diffraction
MCA	Multi-channel analyser
RHEED	Reflection high energy electron diffraction
SEM	Scanning electron microscopy
SIMS	Secondary ion mass spectroscopy
STEM	Scanning transmission electron microscopy
TEM	Transmission electron microscopy
WDS	Wavelength-dispersive spectrometer
XPS	X-ray photon spectrometry (=ESCA)
XRF	X-ray fluorescence
ZAF	Atomic number, absorption and fluorescence corrections

CONTENTS

CHAPTER 1

why electron microscopy?

1.1. *Introduction*

A MICROSCOPE is an optical system which transforms an ' object ' into an ' image '. We are most commonly interested in making the image much larger than the object; in other words, in magnifying the object, and there are very many ways in which we can do this. This book deals with several sophisticated techniques for magnifying images of very small objects by large amounts, but many of the principles involved are just the same as those which have been developed for light microscopes over the past 400 years. The concepts of resolution, magnification, depth of field and lens aberration are very important in electron microscopy and so we deal with them in this first chapter in the more familiar context of the light microscope. When we consider electron microscopes in later chapters it will be found that instead of becoming more complicated many areas of the subject become simpler because we are dealing with electrons rather than light. Thus, although it is apparently more complex, and certainly much more expensive, an electron microscope is almost as easy to understand in principle as its humbler stablemate the magnifying glass. Also, as I hope will be clear by the end of this book, it is a great deal more versatile.

1.2. *The optical microscope*

The simplest optical microscope, which has been in use since the early seventeenth century, is a single lens or ' magnifying glass '. The ray diagram for this is shown in fig. 1.1 and serves to illustrate the concepts of focal length, f, and magnification, m. The image is real but inverted if the object is farther away from the lens than the principal focus (fig. 1.1 (*a*)) and virtual but the same way up if the object is within the focal distance (fig. 1.1 (*b*)). If the image is to be recorded on a photographic plate, then it must be real, and therefore we shall not be concerned with optical arrangements which give rise to virtual images.

Magnification of an object without severe distortion of the image is very limited using a single lens. For higher magnifications, combinations of lenses are used so that the total magnification is achieved in two or more stages. A simple two-stage photomicroscope will have the ray diagram shown in fig. 1.2. The first lens, the ' objective ',

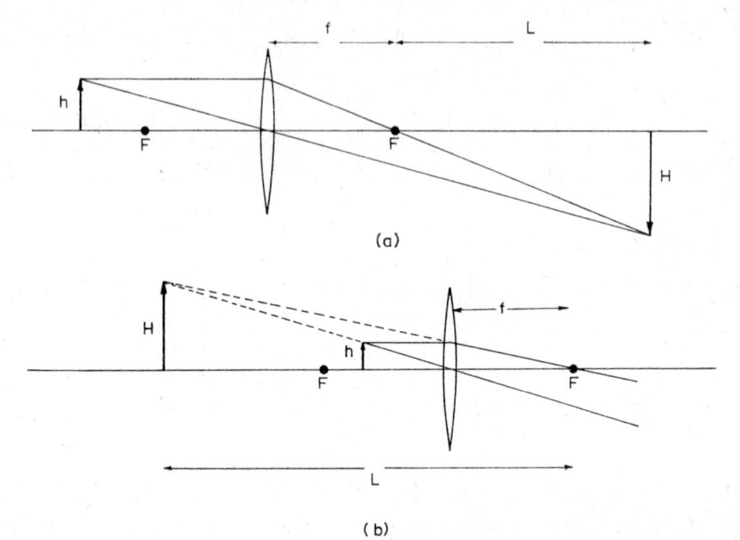

Fig. 1.1. Ray diagrams illustrating the formation of a magnified image by a single lens of focal length f. In both cases the linear magnification H/h is given by L/f.

provides an inverted image at B with a magnification L_1/f_1 and the second lens, the 'projector', gives a final upright image at a further magnification of L_2/f_2. The image is viewed on a screen or recorded on a photographic plate at C with a total magnification of

$$M = L_1/f_1 \times L_2/f_2. \tag{1.1}$$

If higher magnifications are required it is quite straightforward to add a second projector lens to provide a third stage of magnification.

We have so far assumed that the object itself is self-luminous, and we have therefore shown the 'rays' starting at the object and ending up at the viewing screen. In practice we are rarely looking at this sort of specimen and we must illuminate it with light from a convenient source. Now we are forced to consider whether the object

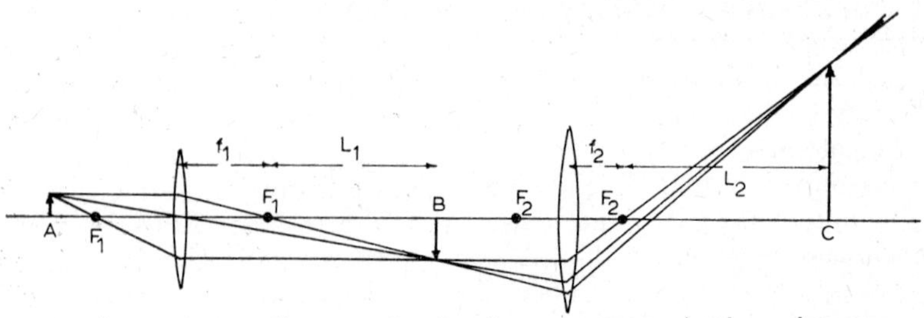

Fig. 1.2. The ray diagram of a simple two-stage projection microscope.

is mainly transparent, in which case we illuminate it from behind, or whether it is opaque, in which case we must illuminate it from in front. Thus, immediately we have a division into two classes of microscope: the biologist who needs to look at very thin sections of tissue uses a transmission arrangement such as is shown in fig. 1.3 (*a*), while the metallurgist or geologist who needs to look at the surface structure of a solid specimen uses a reflection arrangement as shown in fig. 1.3 (*b*). We will see later that the same two types of arrangement arise in electron microscopy.

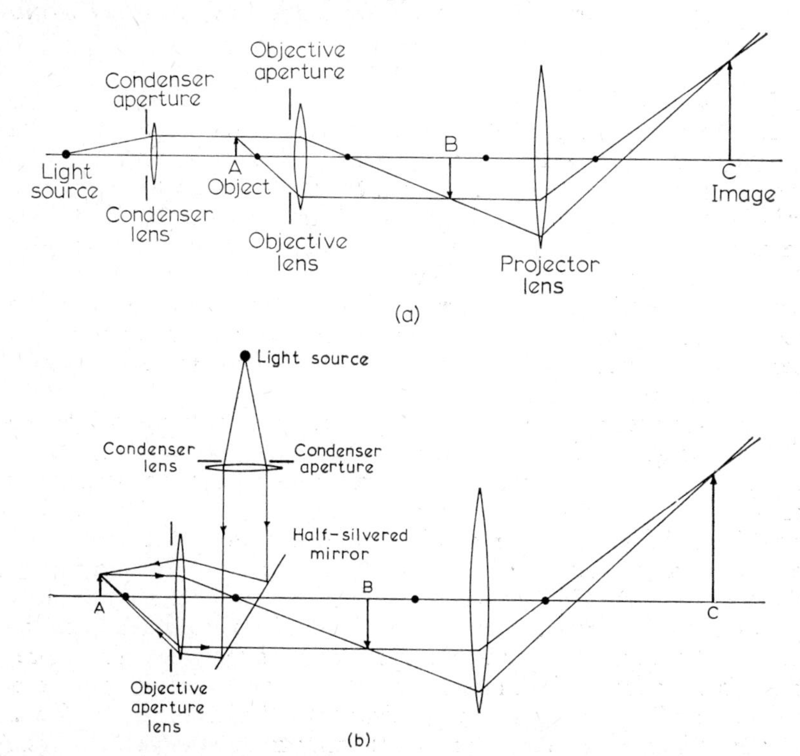

Fig. 1.3. The optical systems for the two common types of projection microscope. (*a*) Transmission illumination; (*b*) reflected illumination.

The essential parts of any illumination system are a light source and a condenser system. The condenser system is necessary to collect the light which is generally diverging from the source and to direct it at the small area of the specimen which is to be examined. This serves two purposes; it makes the object brighter so that we can see it more easily and it also enables us to choose whether to have the illuminating beam of light arriving at the specimen as a converging beam focused on the specimen, or as a parallel beam of light. In early microscopes the sun or ordinary daylight was used as a source and a concave mirror was used to direct the light towards the specimen. For many purposes this is adequate but for more demanding work it is nowadays more usual to find a built-in light source and a condenser lens as is shown in fig. 1.3. With the addition of two variable apertures near the condenser lens and the objective lens we can control the area of specimen which we illuminate and the angular spread of the light collected from the specimen. With a well-made microscope it is then possible to take micrographs such as that shown in fig. 1.4.

1.3. *Magnification*

In principle it is possible to make a light microscope which will give any magnification we care to ask for. However, since for convenience we need a compact instrument without too many adjustments it is usual to alter f_1 or f_2 in equation (1.1) rather than L_1 or L_2. This means that we change one lens for another with a different focal length when we need to change the magnification rather than extending the microscope tube and increasing the distance between the lenses. As a result we can usually choose only a certain number of set magnifications. We will see later that this problem is easily overcome with an electron microscope.

Although we stated in the last section that we could easily increase the total magnification of the microscope by adding additional lenses as additional stages of magnification it turns out that for a vast majority of purposes the two-lens system shown in figs. 1.2 and 1.3 is quite sufficient. The reason for this is simple: the smallest details which can be distinguished in a light microscope are about 200 nm in size (1000 nm = 1 μm, 1000 μm = 1 mm). The reason for this limit is discussed in the next section but for the moment let us consider its implications. The eye can easily detect detail only 0·2 mm in size. Therefore, there is very little point in magnifying the smallest details which can be resolved (200 nm) up to a larger size than 0·2 mm (200 μm). Thus any magnification greater than 1000 × only makes the details *bigger*. We cannot make finer details visible by magnifying the image an extra ten times. An example of this ‘ empty magnification ’, as it is called, is shown in fig. 1.5. The first micrograph has a magnification of 100 × and we can see a lot of detail. Magnifying this ten times more to 1000 × reveals more detail. However, a further stage of magnification to 5000 × shows us no more; we are now looking at features which are further apart, but no clearer. If we need a big display, for instance so that we can view the micrograph at a distance, then it is more sensible to enlarge the 1000 × micrograph photographically than to build a microscope capable of higher magnifications. Now it is

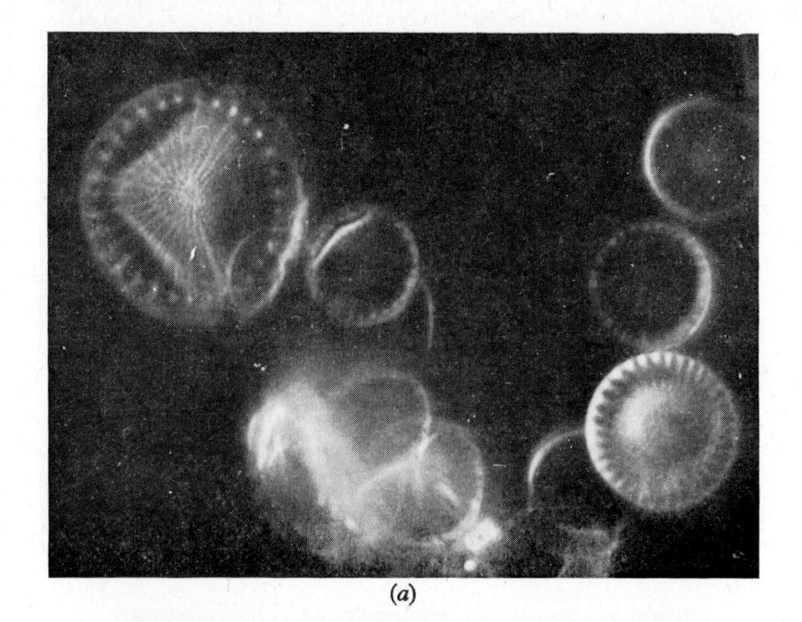

(*a*)

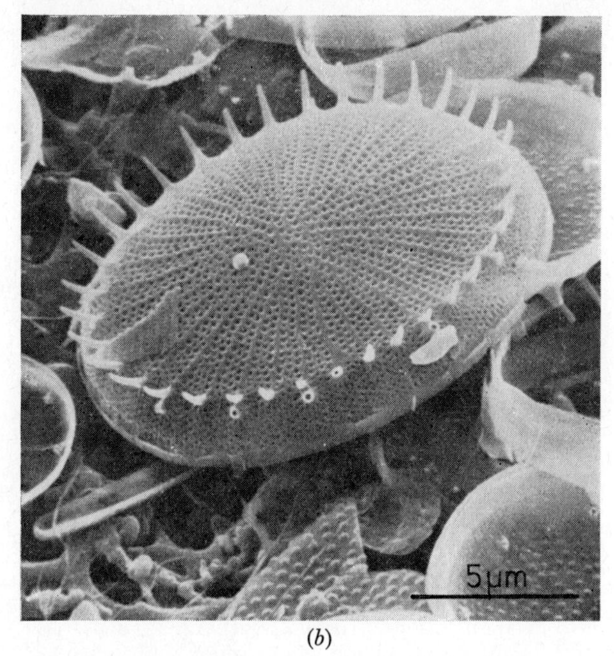

(*b*)

Fig. 1.4. A high resolution optical micrograph (*a*) of a diatom, stephanodiscus (upper left). The limited depth of field is very evident when compared with the scanning electron micrograph of a similar diatom (*b*). (M. O. Moss and G. Gibbs.)

(*a*)

(*b*)

(Fig. 1.5. *Continued on pages* 7 *and* 8)

Fig. 1.5. (*a*), (*c*) and (*e*) Light micrographs of fine pearlite at a range of magnifications. Note that no more detail is visible at the highest magnification than was already clear at a lower magnification. (*b*), (*d*) and (*f*) The same regions as in (*a*), (*c*) and (*e*) photographed in a scanning electron microscope. More detail becomes visible each time the magnification is increased, since the resolution limit of the microscope has not been reached.

(*c*)

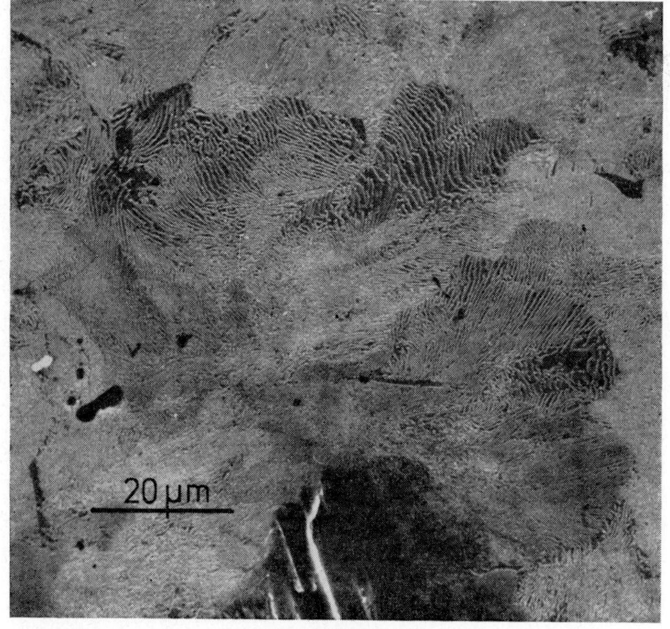

(*d*)

7

(*e*)

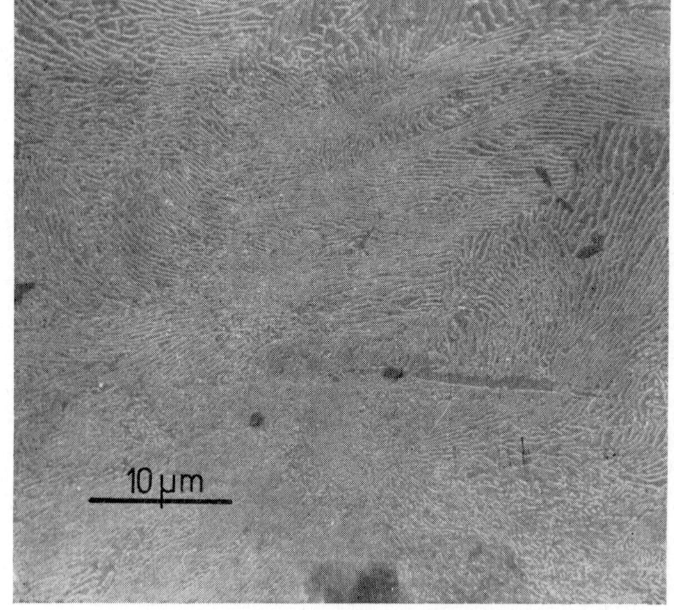

(*f*) (Fig. 1.5. *Caption on page* 6)

relatively easy to provide magnifications of 1000× using only the two-lens system shown in figs. 1.2 and 1.3, for example, using an 80× objective lens and a 15× projector lens. Consequently it is not necessary to build a light microscope with three or more stages of magnification, since this will not improve the *resolution* but rather will degrade it by introducing extra aberrations (see Section 1.6).

1.4. *Resolution*

In order to compare the electron microscope with the light microscope we need to know what factors control the *resolution*, which we can define as the closest spacing of two points which can clearly be seen through the microscope to be separate entities. Notice that this is not necessarily the same as the smallest point which we can see with the microscope, which will often be smaller than the resolution limit.

Even if all the lenses of our microscope were perfect and introduced no distortions into the image the resolution would nevertheless be limited by a diffraction effect. Inevitably in any microscope the light must pass through a series of restricted openings—the lenses themselves or, more likely, the apertures shown in fig. 1.3. Wherever light passes through an aperture diffraction occurs so that a parallel beam of light, which we would see as a spot, is transformed into a series of cones, which we see as circles, known as Airy's rings. Figure 1.6 shows this effect with a laser beam and two small pinholes. For light of a given wavelength the diameter of the central spot is inversely proportional to the diameter of the aperture from which the diffraction is occurring. Consequently the smaller the aperture the larger is the central spot of the Airy disc. We have used very small apertures in order to make the Airy disc clearly visible but the same effects occur from the relatively large apertures found in light microscopes. The diffraction effect limits the resolution of a microscope, since the light from every small point in the object which we would like to make visible suffers diffraction, particularly by the objective aperture, and even an infinitely small point becomes a small Airy disc in the image. In order to make this disc as small as possible, in other words, to make the image of a point as small as possible, we must use as large an aperture as we can.

Now let us consider the resolution of the microscope in more detail. First we need to know more about the Airy disc. If we plot the intensity of light as we cross the series of rings which make up the disc, we find a distribution similar to that shown in fig. 1.7; the central spot is very much more intense than any other ring and in fact contains 84% of all the light intensity. Consequently for many purposes we can ignore the rings and assume that all the light falls in a spot of diameter d_1, where d_1 is $\propto 1/(\text{aperture diameter})$. We next have to consider how far apart two of these spots in the image must be before

(*a*)

(*b*)

Fig. 1.6. Airy rings formed by the diffraction of a laser beam at (*a*) a 50 μm pinhole and (*b*) a 100 μm pinhole. The diameter of the central bright spot is larger from a smaller aperture.

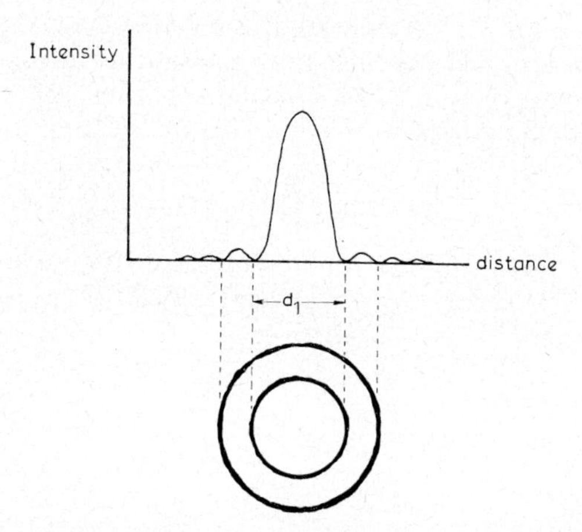

Fig. 1.7. The variation of light intensity across a set of Airy rings. Most of the light (84%) falls inside the first dark ring, that is within a spot of diameter d_1.

we can tell that there are two of them. This distance is the resolution which we defined earlier. Lord Rayleigh proposed a criterion which works well in most cases and has been used extensively ever since: when the maximum of intensity of an Airy disc coincides with the first minimum of the second, then the two points can just be distinguished. This is illustrated in fig. 1.8, from which it can be seen that the

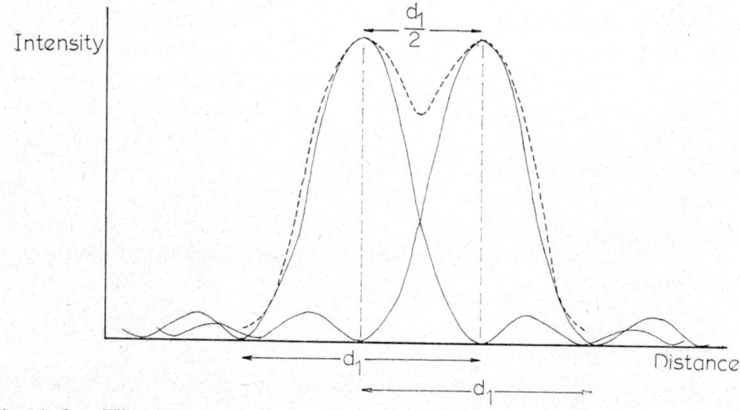

Fig. 1.8. The intensity of the Airy rings from two neighbouring pinholes. The intensity distributions from each of the pinholes separately are shown as solid lines; the combined profile from the two pinholes acting together is shown dotted. At the Rayleigh resolution limit, as shown here, the maximum intensity from one pinhole coincides with the first minimum from the other. This gives a resolution limit of $d_1/2$.

resolution limit is $d_1/2$. We normally refer to microscope apertures in terms of the semi-angles which they subtend at the specimen (α in fig. 1.9). It is then possible to derive from diffraction theory (Ditchburn, 1963; Born and Wolf, 1970) the relationship

$$d_1 = \frac{0{\cdot}61\,\lambda}{\mu \sin \alpha}, \tag{1.2}$$

where λ is the wavelength of the light and μ is the refractive index of the medium between the object and the objective lens.

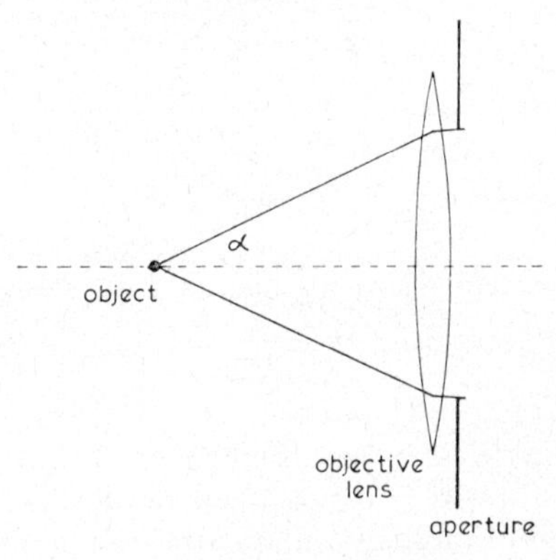

Fig. 1.9. The definition of the half-angle, α, subtended by an aperture (in this case the objective aperture).

In order to obtain the best resolution (i.e. smallest d_1) we obviously need to decrease λ or increase μ and α. With a light microscope we can decrease λ to ~400 nm by using green light (or to ~200 nm if it is possible to use ultraviolet light); we can increase sin α towards 1 by using as large an aperture as possible and we can increase μ by using an oil immersion objective. However, it is impractical to make $\mu \sin \alpha$ much greater than 1·60 since sin α must be less than unity and even very exotic materials are limited to a refractive index of about 1·7. The absolute resolution limit using green light is therefore about 150 nm.

1.5. *Depth of field*

In any microscope the image is only accurately in focus when the object lies in the appropriate plane (strictly the surface of a sphere). If part of the object being viewed lies above or below this plane, then the equivalent part of the image will be out of focus by some amount. The distance above and below the object's focal position for which our eye can detect no change in the sharpness of the image is known as the depth of field. In most microscopes this distance is rather small and therefore in order to produce sharp images we must start with very flat objects. If we try to view an object which is not flat at high magnification using a light microscope then we must tolerate some out-of-focus regions, as is illustrated in fig. 1.4. This is a useful feature if we wish to accentuate certain parts of the image at the expense of others, but is a grave disadvantage if we would like to see all parts of a three-dimensional object clearly.

The depth of field of a particular microscope depends critically on its resolution and on the magnification to be employed: as the resolution is improved and the magnification is raised the depth of field becomes smaller. A fairly accurate expression for the depth of field, Δ (in millimetres), of a microscope is

$$\Delta = \frac{\lambda}{\mu \sin^2 \alpha} + \frac{1}{7m \sin \alpha}, \tag{1.3}$$

where m is the magnification.

Since λ is $\sim 0{\cdot}0004$ mm for high resolution the second term is the larger at low magnification, whereas in conditions of high magnification the two terms are of similar size and will limit the depth of field to less than 1 μm. There is nothing that can be done to improve the depth of field in order to examine specimens with three-dimensional character unless we are prepared to sacrifice resolution by decreasing α and working at low magnifications. Thus if we need to examine a specimen at 1000 × we shall not see clearly details deeper than ~ 1 μm, whereas at 40 × we may be able to tolerate variations of 100 μm in the specimen surface. Unfortunately many of the real specimens we would like to study have details with a greater depth than a tenth of a millimetre. Try to focus on a fly at 40 × !

1.6. *Aberrations in optical systems*

While discussing resolution and depth of field we have assumed that all the components of our microscope are perfect and will focus the light from a point on the object to a similar unique point in the image. This is in fact rather difficult to achieve because of lens aberrations. The easiest lenses to make are those with spherical surfaces but any single spherical lens suffers from two types of aberration—*chromatic* aberrations which depend on the wavelength content of the light and *monochromatic* aberrations which affect even light of a single wavelength. The effect of each aberration is to distort the image of every point in the object in a particular way, leading to an overall loss of quality and resolution in the image. In order to correct these aberrations it is necessary to replace the single lenses we have shown in our diagrams with compound lenses containing several carefully shaped pieces of glass with different refractive indices. Although this is not a correction technique which we can use in the electron microscope the same types of aberration arise (and are very important in determining the resolution of the instrument) and therefore we need to consider the most significant aberrations in more detail.

Chromatic aberrations occur when a range of wavelengths is present in the light (e.g. sunlight) and arise because a single lens causes light to be deviated by an amount depending on its wavelength. Thus a lens will have different focal lengths for light of different wavelengths. To take extremes a red focus and a blue focus will be formed from white light. Figure 1.10 makes this clear in terms of ray paths and illustrates that wherever we view the image we will see coloured haloes surrounding each detail. For example, if we place the viewing screen at A we shall see the image of a point as a bluish dot with a red halo, whereas if the screen were at B we would see a reddish dot with a blue halo. In neither case do we see a truly focused image of a small white dot. With the screen at the compromise position C we have the smallest dot image but instead of a point it is a *circle of least confusion*.

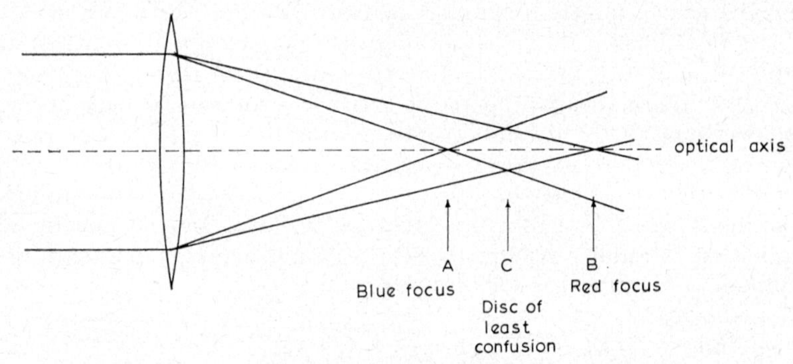

Fig. 1.10. Ray diagram illustrating the introduction of chromatic aberration by a single lens. Light of shorter wavelength (blue) is brought to a focus nearer the lens than the longer wavelength red light. The smallest 'focused' spot is the disc of least confusion at C.

All aberration corrections are designed to reduce in size this circle of confusion. In the light microscope there are two ways in which we can improve the chromatic aberrations; either we can combine lenses of different shapes and refractive indices or we can eliminate the variation in wavelength from the light source by the use of filters or special lamps. Both methods are often used if the very best resolution is required.

Monochromatic aberrations arise because of the different path lengths of different rays from an object point to the image point.

The simplest of these effects is *spherical aberration*, which is illustrated in fig. 1.11. The portion of the lens farthest from the optical axis brings rays to a focus nearer the lens than does the central portion of the lens. Another way of expressing this concept is to say that the optical ray path length from object point to focused image point should always be the same. This naturally implies that the focus for marginal rays is nearer to the lens than the focus for axial rays. Again a circle of least confusion exists at the best compromise position of focus.

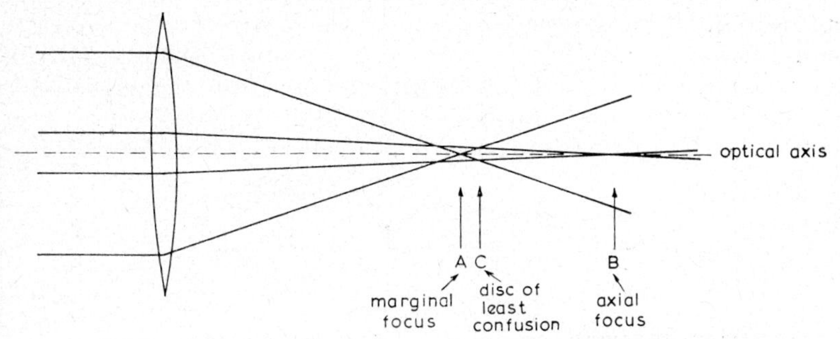

Fig. 1.11. Ray diagram illustrating spherical aberration. Marginal rays are brought to focus nearer the lens than near-axial rays.

A related effect is that of astigmatism. This is illustrated in fig. 1.12. For object points off the optical axis the path length criterion shows that there will be a focus for horizontal rays at a different position from the focus for vertical rays. A similar effect arises if the lens does not have identical optical properties across the

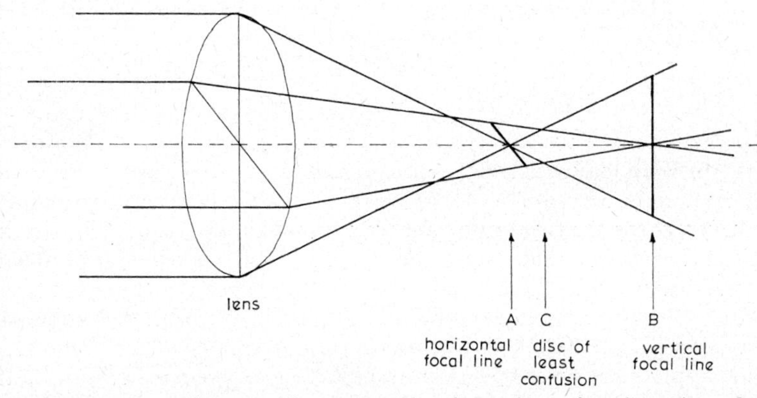

Fig. 1.12. Ray diagram illustrating the formation of astigmatism for a lens with slightly different optical properties in the horizontal and vertical directions. In this illustration the lens is more powerful in the vertical plane.

whole of its area. All the monochromatic aberrations are improved if only the central portion of the lens is used, i.e. if the lens aperture is 'stopped down'. Unfortunately, as we see in Section 4, this limits the resolution of the microscope.

Other aberrations are often discussed in textbooks on optics but the three mentioned here are the three of prime concern in electron microscopy. One further effect which is sometimes troublesome, particularly at low magnifications, is distortion. This occurs when for some reason the magnification of the lens changes for rays off the optical axis. The two possible cases are when magnification increases with distance from the optical axis, leading to pin-cushion distortion, and when magnification decreases with distance from the optical axis, leading to barrel distortion (fig. 1.13). This is obviously of great importance if measurements are to be made from micrographs.

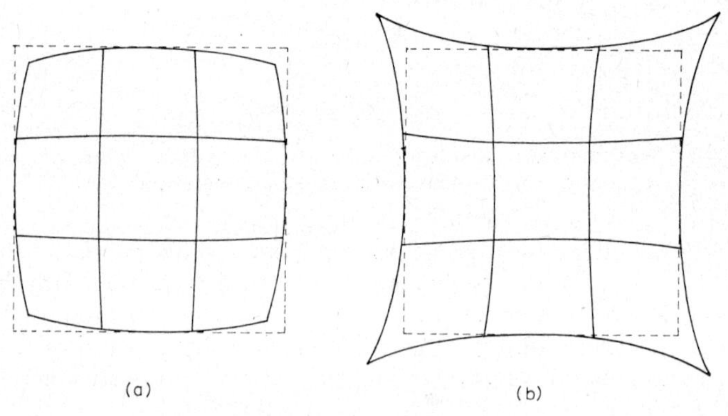

Fig. 1.13. The appearance of a square grid in the presence of (*a*) barrel and (*b*) pincushion distortion.

1.7. *The effect of using electrons instead of light*

In very many ways electron optics is just the same as light optics—all the terminology we have been using above applies and we can still use ray diagrams to illustrate the working of the electron microscope. However, there are two important ways in which electrons differ from light and each has led to the development of a particular type of electron microscope.

Most of us are used to thinking of light as electromagnetic radiation with a wavelength λ and of electrons as sub-atomic particles. We can in fact apply both types of description (*wave* and *particle*) to both light and electrons: thus light may be described in terms of photons or as radiation of wavelength 400–700 nm while electrons can also be considered as radiation with wavelengths (useful in microscopy) between about 0·001 nm and 0·01 nm. The first obvious difference between electrons and light is that their wavelengths differ by a factor of many thousands. The implications of this for microscopy are enormous but fortunately in most cases lead to a simplification!

The most obvious difference arising from the shorter wavelength of an electron beam is that electrons are very much more easily absorbed by many materials than is light. This is so severe an effect that in order to use electrons in a microscope we must evacuate all the optical paths to a pressure of better than 10^{-4} Torr ($\times 10^{-7}$ of atmospheric pressure); the electrons would scarcely penetrate a few millimetres of air at atmospheric pressure. Since, as we shall see in the next chapter, the lenses in an electron microscope are merely magnetic fields we have only small changes of refractive index as the electrons pass through each lens. Hence in all our calculations we can write $\mu = 1$. Furthermore, the angles through which the 'rays' need to be deflected are generally very small (a few degrees) and we can make the approximation $\sin \alpha = \alpha$ (α in radians) (fig. 1.9) to a very high degree of accuracy. These simplifications mean that we can write the theoretical resolution of the electron microscope (equation (1.2)) as

$$d_1 = \frac{0{\cdot}61\lambda}{\alpha} \qquad (1.4)$$

which should mean that with reasonable values of $\lambda = 0{\cdot}0037$ nm and $\alpha = 0{\cdot}1$ radians we expect a resolution of about 0·02 nm, which is much smaller than a single atom. Unfortunately, however, in the *transmission electron microscope* (TEM) we cannot obtain this sort of resolution because of the lens aberrations. Whereas in a light microscope it is possible to correct both chromatic and monochromatic aberrations by using subtle combinations of lenses, there is no such possibility with an electron lens since one cannot change its refractive index. Consequently although chromatic aberrations can be virtually eliminated by using electrons of a very small range of wavelengths it is not possible to eliminate the monochromatic aberrations, principally spherical aberration. The only way of minimizing this is to restrict the electrons to paths very near the optical axis, i.e. near the centre of the lens, by using a small objective aperture. This reduces α and therefore makes the resolution worse. There is an optimum size of aperture (i.e. value of α) for which the resolution is smallest; using this aperture it is nowadays possible with a good transmission electron microscope to resolve two points about 0·3 nm apart. This is approximately the separation of atoms in a solid.

Since it is necessary to keep α small in order to reduce the effect of spherical aberrations we also have the benefit of a larger depth of field when using electrons. Equation (1.3) rewritten to include the electron approximation as

$$\Delta = \frac{\lambda}{\alpha^2} + \frac{1}{7m\alpha}\ \text{mm}, \qquad (1.5)$$

shows that as α is reduced the depth of field increases very rapidly. This is one of the major advantages of electron microscopes.

The second major difference between electrons and light is that electrons carry a charge. Not only does this mean that we can use electromagnetic fields as lenses for electrons but it opens up the possibility of scanning a beam of electrons back and forth, as happens in a cathode-ray tube or a television tube. The application of this approach has led to the development of the *scanning electron microscope* (SEM) which, as we shall see in Chapter 4, has in the past ten years revolutionized our attitude to the study of surfaces.

With both types of electron microscope, transmission and scanning, the use of electromagnetic lenses and deflection coils means that it is possible to obtain an image of the specimen at any magnification within a wide range (say up to 100 000 ×) without changing lenses. Electron microscopes therefore offer higher resolution, higher magnification, greater depth of field and greater versatility than the light microscope, although at a rather higher price (tens of thousands of pounds).

CHAPTER 2

electrons and their interaction with the specimen

2.1. *Introduction*

WHEN thinking about light microscopy we tend to ignore most of the interactions between the light and the specimen. We are concerned only that enough light is transmitted through or reflected from the specimen so that we can easily see the image and we assume that the specimen is unchanged by the fact that we have looked at it. For most specimens this is a fair assumption. However, when we are dealing with electrons, their interaction with the material through which they pass may have more serious consequences; for example, we have already seen that because of the scattering of electrons by gas molecules in the air we must operate the electron microscope in a vacuum. Some of the other very real possibilities are that the specimen will be heated by the electron beam and that chemical changes may take place.

It is clearly important, if we are to appreciate the way in which an electron microscope works and the meaning of the information which it gives us, that we understand the nature of the possible interactions between the electron beam and the other parts of the microscope (e.g. the lenses and the camera) and between the electrons and the specimen. In order to achieve this understanding we must consider in more detail the nature of the electron and the various possible interactions between an electron and an atom. Those readers who need only a superficial understanding of electron microscopy or who wish to get the feel of the subject on a first reading may like to skip the rest of this chapter and pass on to Chapter 3.

2.2. *Properties of electrons*

Two schematic ways of looking at the structure of a typical atom are shown in fig. 2.1. The nucleus carries a large positive charge and is surrounded by a cloud of negative electrons which exactly neutralize this charge. For most purposes we can consider that the electrons tend to occupy particular orbits or ' shells ', which were long ago labelled K, L, M and so on, as shown in fig. 2.1 (*a*). A very light element, such as helium, which has only two electrons, will have electrons in its K shell only and the L, M and higher shells will be empty. Uranium, with an atomic number of 92 and hence 92 electrons, will have electrons in its K, L, M, N, O, P and Q shells! The electrons nearest to the nucleus are those which are removed from the atom with the greatest difficulty while those in the outermost shells can be removed if they are given only a very little extra energy. We can represent this on an energy level diagram such as fig. 2.1 (*b*). Not all the electrons in each shell have exactly the same energy; this will affect our discussion of X-rays in Chapter 5. The alternative nomenclature for describing electron orbits (1s, 2s, 2p, etc.) is also indicated on fig. 2.1 (*b*).

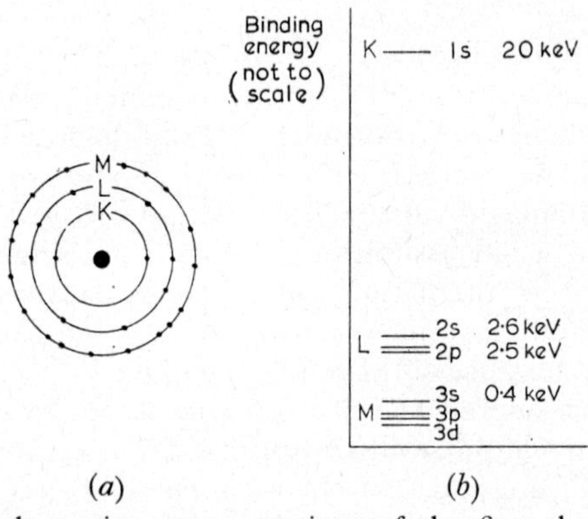

Fig. 2.1. Two alternative representations of the first three electron shells around the nucleus of a molybdenum atom. The innermost shell (K) is the most tightly bound and therefore the two K electrons need to receive the largest amount of energy (20 keV) before they can be ejected from the atom.

Electrons are fairly readily detached from their atoms, since we can easily supply the necessary energy, as we shall see later. It is this easy availability and their low mass which makes them so useful, but before we can predict how free electrons will behave in an electron microscope we need to know a bit more about them. The rest of this section therefore considers the mass, velocity and wavelength of those electrons with sufficient energy to be used in electron microscopy.

An electron, considered as a particle, carries a single negative charge, e, measured as about $1{\cdot}6\times10^{-19}$ C and has a rest-mass m_e about 9×10^{-31} kg. If a single electron is accelerated by a large potential difference V, then its velocity, v, may well approach the velocity of light, c, and relativistic effects will become important. One of these is that its mass will increase compared with its *rest-mass*, m_e, according to the equation

$$m=\frac{m_e}{\sqrt{\left(1-\left(\frac{v}{c}\right)^2\right)}}. \tag{2.1}$$

If we now think of the electron in terms of a wave, then its wavelength and its momentum must be connected by de Broglie's relationship

$$\lambda = \frac{h}{mv}, \tag{2.2}$$

where h is the Planck constant. In addition we can equate the energy given to the electron, eV, with the energy represented by the relativistic change in mass

$$eV = (m - m_e)c^2. \tag{2.3}$$

If we combine equations (2.1), (2.2) and (2.3) we find that the wavelength of the electron depends on the potential difference, or *accelerating voltage* in the following way:

$$\lambda^2 = h^2/(2eVm_e + e^2V^2/c^2)$$

which when we substitute the values of h, e, m_e and c becomes

$$\lambda = \sqrt{\left(\frac{1{\cdot}5}{V + 10^{-6}V^2}\right)}\,\text{nm}. \tag{2.4}$$

At the accelerating voltages which we find useful for electron microscopy (2×10^4 V upwards) the electrons are accelerated to a velocity which is a significant fraction of the velocity of light and the relativistic effects are quite significant. Consequently, the wavelength must be calculated according to equation (2.4) and not according to the simpler expression

$$\lambda = \sqrt{\left(\frac{1{\cdot}5}{V}\right)}\,\text{nm}.$$

Table 2.1

V/kV	$\lambda \propto V^{-1/2}$	λ relativistically corrected (nm)
20	0·0086	0·0086
40	0·0061	0·0060
60	0·0050	0·0049
80	0·0043	0·0042
100	0·0039	0·0037
200	0·0027	0·0025
500	0·0017	0·0014
1000	0·0012	0·0009

$c = 2{\cdot}998 \times 10^8$ m s^{-1}
$e = 1{\cdot}602 \times 10^{-19}$ C
$h = 6{\cdot}62 \times 10^{-34}$ J s
$m_e = 9{\cdot}108 \times 10^{-31}$ kg

As table 2.1 shows, the effect of relativity is quite significant for high accelerating voltages, amounting to a 25% correction at one million volts.

2.3. *Generating a beam of electrons*

Of the many ways of encouraging electrons to leave a metal so that they may be accelerated towards the specimen, two have proved useful in the construction of *electron guns*. The most widespread system uses thermionic emission from a heated tungsten filament. At temperatures in excess of 2700 K a tungsten wire emits an abundance of both light and electrons; in a light bulb we use only the light but in an electron gun we also accelerate the electrons across a potential difference of tens or hundreds of kilovolts to generate a beam of electrons of known energy (and hence of known velocity and wavelength). The general features of a thermionic triode electron gun are shown in fig. 2.2. A piece of tungsten, usually a wire bent into a hairpin, acts as a cathode. This filament (F) is heated by the passage of a current to about 2800 K while being held at a high negative potential with respect to the anode (A) and the rest of the microscope. Electrons emitted from the filament are accelerated rapidly towards the anode and a beam of high energy electrons is emitted through the circular hole at its centre into the microscope column. The addition of a Wehnelt cap (W), which is held at a voltage slightly more negative than the filament, enables the diameter of the area at the end of the filament which emits electrons to be controlled. The Wehnelt cap thus acts rather like the grid in a triode valve and hence this gun is usually called a triode gun. The most important feature of the gun is that the paths of the electrons leaving the anode usually cross at one point in space and hence the gun is acting as a lens. The diameter of the beam at the crossover is dependent on the area of the filament which is emitting electrons and hence can be controlled by the difference in potential between the filament and the grid (i.e. by R_b, the resistance of the bias resistor in fig. 2.2). This crossover diameter, analogous to the circle of least confusion which was defined in Chapter 1, is effectively the size of the electron source, and is of great importance in calculations concerning the resolution of electron microscopes.

The thermionic gun is satisfactory for the majority of purposes but is limited in the *brightness* of the beam that it can emit. In the context of electron microscopy we define brightness as the current density per unit solid angle (e.g. in A mm^{-2} sr^{-1})†; in other words

† sr = steradian, the unit of solid angle. One steradian is defined as that solid angle which encloses a surface on a sphere equal to the square of its radius. 4π steradians therefore occupy the whole of space.

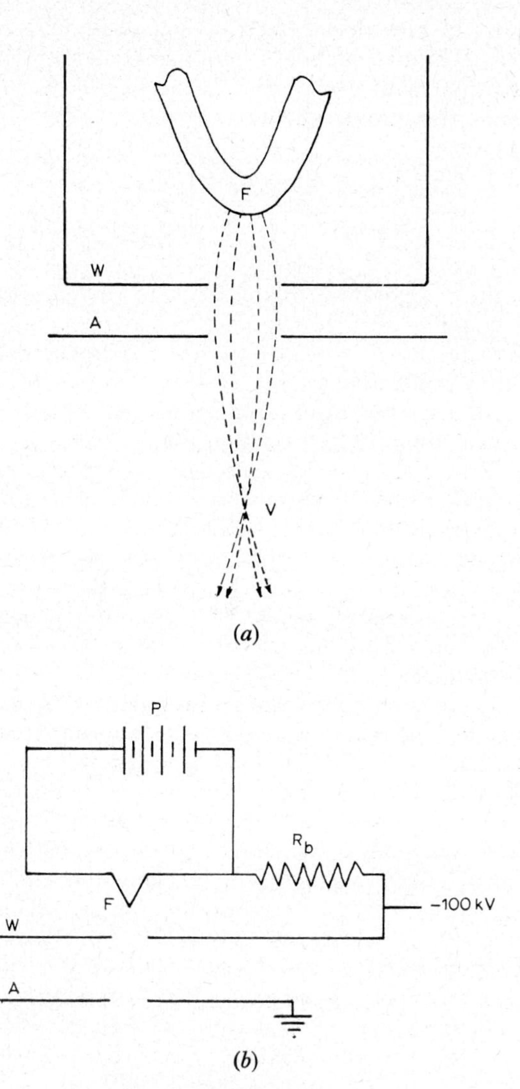

Fig. 2.2. The geometry and electrical layout of a thermionic triode electron gun. (*a*) The electrons are emitted from a small region at the tip of a heated tungsten filament (F) and are accelerated towards the anode (A). The fields generated between the filament and the anode, modified by the Wehnelt cylinder (W) acting as a grid, cause the electrons to be focused at V, which is known as the virtual source. (*b*) The filament is heated by the passage of a current from the power supply P and the voltage on the grid is determined by the bias resistor R_b.

it is a measure of how many electrons per second we can pass along a given electron beam and hence how many electrons per second we can direct to a particular small area of the specimen. Brightness, B, is limited for a thermionic gun to

$$B = \frac{1{\cdot}2TV \exp(-\phi/kT)}{\pi k} \text{ A mm}^{-2}\text{ sr}^{-1}, \tag{2.5}$$

where T is the temperature of the filament (in degrees Kelvin) and ϕ is the thermionic work function of the filament in electron volts. Since B increases as T increases we use a filament material with as high a melting point and as low a work function as is practicable. Tungsten is almost universally used (m.p. $=3653$ K, $\phi=4{\cdot}52$ eV) and the brightness is then limited to about 2×10^3 A mm^{-2} sr^{-1}. This is sufficiently bright for most conventional transmission electron microscopy but greater brightness is needed to obtain micrographs at very high resolution using the scanning transmission electron microscope (see Chapter 4). Since brightnesses greater than those given by equation (2.5) cannot be produced using a thermionic electron gun, an alternative method of generating the electrons must be used. The only available method employs *field emission*, which we will now consider.

It a metal surface is subjected to an extremely high electric field (greater than 10^9 V/m) there is a very good chance that an electron can leave the surface without needing to be given the amount of energy represented by the work function. This is because the effect predicted by quantum mechanics and known as ‘ tunnelling ’ can occur. The result is that many more electrons can be drawn from a piece of tungsten than is possible using thermionic emission, and the brightness can be increased by a factor of about one million. Unfortunately in order to make such a field emission gun work several rather stringent experimental conditions must be met. The only reasonable way of creating an electric field greater than 10^9 V/m is to make the tip of the tungsten filament extremely sharp—the diameter at the point must be about $0{\cdot}1\ \mu$m, which makes the filament extremely fragile (the diameter of the point of a pin is about $100\ \mu$m). The other main difficulty is that the vacuum in the gun must be better than 10^{-9} Torr compared with 10^{-4} or 10^{-5} Torr for a thermionic gun. This means that very expensive techniques, which are at present likely to be fairly unreliable, must be used. On the positive side the field emission mechanism is hardly dependent at all on the temperature of the filament and so the tungsten point can be used at room temperature.

2.4. *Deflection of electrons—lenses*

It was realized in the 1920's that a beam of electrons could be focused by either an electrostatic field or a magnetic field. Both types of field have been used in electron lenses but the electromagnetic lens is by now virtually universal so we will not consider the earlier electrostatic lenses any further.

The key to an understanding of what is essentially a very simple lens is the direction of the force which acts on a moving electron in a magnetic field. If an electron moving with velocity v experiences a magnetic field of strength B, then it suffers a force $F = Bev$ in a direction perpendicular to both the direction of motion and the magnetic field. Expressed more concisely in vector notation:

$$\boldsymbol{F} = e(\boldsymbol{B} \wedge \boldsymbol{v}).$$

Now a typical electromagnetic lens is designed to provide a magnetic field almost parallel to the direction of travel of the electrons. An electron entering the lens (fig. 2.3) experiences a magnetic field B which we can conveniently resolve into components B_{ax} along the axis of the microscope and B_{rad} in a radial direction. Initially the electron is unaffected by B_{ax}, which is parallel to its direction of travel, but experiences a small force $B_{\text{rad}}ev$ from the small radial component. This force causes the electron to travel in a helix along the lens. As soon as it starts to spiral, however, it has a component of velocity $v_{\perp}$ perpendicular to the plane of the paper and therefore experiences a

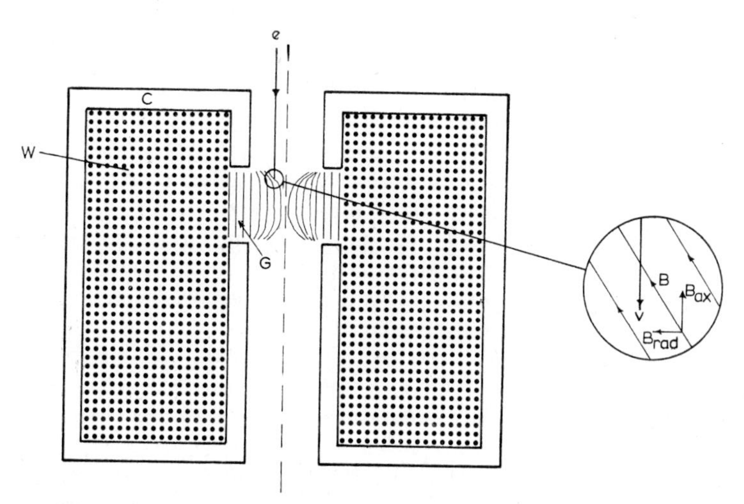

Fig. 2.3. The action of a typical electron lens. An electron (e) entering the lens just off the optical axis experiences a magnetic field such as that shown in the inset.

force $B_{ax}ev_{\perp}$ in a radial direction. Thus the helical path follows a tighter and tighter radius and the effect is that a parallel beam of electrons entering the lens is caused to converge to a point exactly as light is focused by a glass lens. If the magnetic field only extends over a very short distance along the axis, then the lens behaves as a 'thin lens' and all the geometrical expressions quoted in Chapter 1 apply.

In order to generate a magnetic field of just the right strength, size and shape, an arrangement similar to that shown in fig. 2.3 is almost invariably used. A coil consisting of a large number of turns of wire, W, is wound on a soft iron core, C, which has only a very small accurately machined air gap, G, across which the field is produced. By varying the current passing through the coil, typically from zero to about one ampere, the magnetic field strength and hence the focal length of the lens can be varied at will.

Although the several lenses in any one electron microscope may differ in shape and size they will conform to the general pattern laid out in fig. 2.3. An important feature, for which there is no analogy in the light microscope, is the spiralling of the electrons as they travel through an electromagnetic lens. Since it is very rare for the electron to travel an integral number of turns of the spiral as it passes through the lens, in general there is a rotation of the image caused by the lens. This is not a distortion, since the image is otherwise unaffected, but it does lead to one or two effects, particularly in transmission electron microscopes, which we shall need to bear in mind when looking at micrographs and electron diffraction patterns in later chapters.

2.5. *The scattering of electrons by the atoms of the specimen*

In all types of electron microscope, except one which we shall not consider, electrons enter the specimen and the same or different electrons leave it again to form the image. Consequently it is vitally important that we understand the interactions which are possible between electrons and atoms in a solid specimen so that we can interpret the image. There are really only four possibilities for each incoming electron, illustrated in fig. 2.4, but we need to know how likely each one is for a particular specimen.

The first, and least likely, possibility is that the incident electron passes straight through the specimen without interacting with it in any way. This will only occur if the specimen is very thin; $t \ll 1$ μm for most materials.

The second possibility is that the electron passes very close to the strongly positively charged nucleus of an atom and is deflected from its path by the attraction of the opposite charges. Depending on how close the electron approaches to the nucleus and how fast it is travelling (i.e. its energy) it may be deflected only slightly from its path or at the other extreme may be deflected almost 180° and travel back in the direction from which it came. In the majority of cases the angle of scattering, θ, lies between these two extremes. The significant point about this type of scattering is that the electron loses virtually no energy in the process: it changes its direction but not its energy and therefore is not slowed down. This type of scattering is called *elastic* scattering and for most materials will be the most common occurrence for the incident electrons.

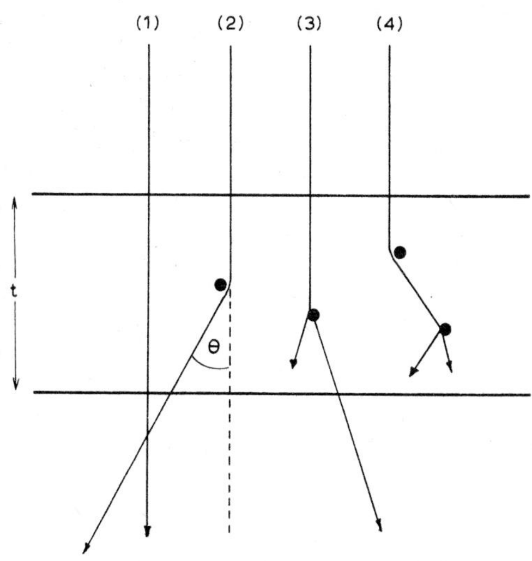

Fig. 2.4. The four possible types of interaction which an electron can have with a specimen of thickness t. (1) No scattering of any sort; (2) elastic scattering by a single atom, causing a deflection θ; (3) inelastic scattering by a single atom, resulting in two slower electrons; (4) both elastic and inelastic scattering, perhaps several times if the specimen is fairly thick.

The third possibility is that the electron interacts with one of the orbital electrons of an atom. In this case the two identical charged particles repel one another and some of the energy of the incident electron is transferred to the orbital electron. Both electrons then generally move on, leaving a vacant electron site in one of the shells of the atom. The incident electron has lost energy and is therefore slowed down, as well as being changed in direction, so this type of scattering is known as *inelastic* scattering. If one electron suffers a succession of inelastic collisions it will eventually lose all its energy and will be stopped in the specimen. It is then effectively *absorbed* and the probability of this happening before the electron reaches the other side of the specimen obviously increases with the thickness of the specimen. This is why, in specimens thicker than 1 μm, very few electrons are *transmitted* right through the specimen; the absorption, in fact, follows similar laws to the absorption of light or X-rays in a solid, the enormous differences being in the value of the absorption coefficient μ in an expression of the form

$$I = I_0 \exp(-\mu t), \tag{2.6}$$

where I is the intensity transmitted through a specimen of thickness t and I_0 is the incident intensity. (This expression is known as Lambert's law when applied to the absorption of light in a liquid solution.)

An important feature of inelastic scattering is that it is 'peaked' in the forward direction; in other words, the greater the angle of deflection of the incident electron the smaller is the chance of that deflection occurring. We can summarize the behaviour of 25 electrons in fig. 2.5. If each of the 25 is incident on the specimen along AB, then their paths after inelastic scattering are likely to be those shown in the figure, with only a few scattered through large angles. We shall meet one consequence of this distribution when we discuss Kikuchi lines in electron diffraction patterns in Chapter 3.

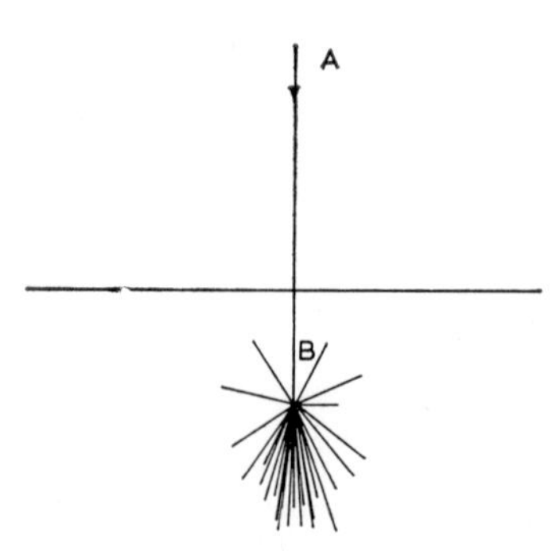

Fig. 2.5. A schematic representation of the trajectories of 25 electrons which have been inelastically scattered at B. The majority of electrons are not deflected very much from their initial direction AB.

The remaining possibility illustrated in fig. 2.3 for an incoming electron is perhaps the most likely of all, as long as the specimen is thicker than about 20 nm. This is that the incident electron will be both elastically *and* inelastically scattered, probably many times. The result is that each electron follows a zig-zag path such as shown in fig. 2.6 before finally coming to rest or reaching the other side of the specimen. Each time the incident electron is inelastically scattered it knocks an electron out of its orbit in an atom of the specimen and leaves the atom in an *excited state*. There are many ways in which the atom can relax from its excited state and although all of them finally result in an electron filling the gap in the orbital, the energy given off as this happens manifests itself in a variety of ways. These lead to the so-called secondary effects which can give us much information about the composition and nature of the specimen. We will consider them individually in the next section.

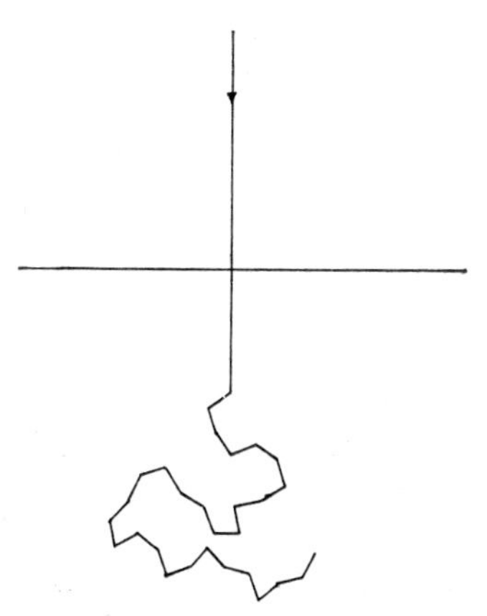

Fig. 2.6. A typical trajectory for a single electron as it is scattered many times after entering a solid specimen.

2.6. *Secondary effects generated by the incident electron beam*

Some of the secondary effects which can be generated in a specimen and detected outside it are summarized in fig. 2.7. The first effect we have almost come across already. The electrons which are knocked out of their orbits around particular atoms are given a small amount of energy by the incoming (primary) electron. The majority of these electrons will wander about the specimen for a short time before being accepted by other excited atoms which have electrons missing from one of their orbitals. However, if the electron is near the surface of the specimen (within about 20 nm) it may have enough energy to escape from the specimen and will become what we call a secondary electron. When it leaves the specimen it will have a very low energy (and hence velocity), usually less than 100 eV, compared with the incident primary electrons, usually several thousand electron volts. We use these secondary electrons to form the image in the scanning electron microscope (Chapter 4).

A second type of effect occurs when an electron fills the vacant electron site in one of the orbitals of an excited atom. This means

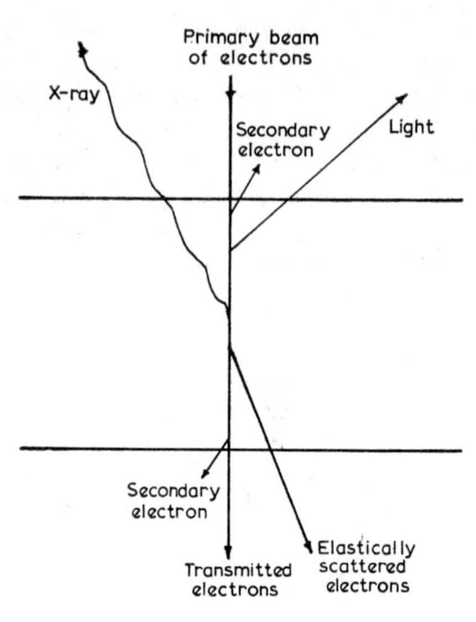

Fig. 2.7. A summary of the secondary effects which may be excited when a primary beam of electrons hits a specimen.

that the atom changes from a high-energy state (excited) to a low-energy state (normal stable state) and therefore some energy has to be given out. One of the ways in which this can happen is for an X-ray of the appropriate energy to be emitted. The X-ray will have a wavelength, λ, depending on the difference in energy, ΔE, between the two states of the atom, according to

$$\lambda = \frac{hc}{\Delta E}, \tag{2.7}$$

where h is the Planck constant and c the velocity of light. For example, consider that one of the electrons had been knocked out of the K shell of the molybdenum atom shown in fig. 2.1 (*a*). If an electron from the L shell jumped in to fill the vacancy, then ΔE would be 17 400 eV (from fig. 2.1 (*b*)) and the wavelength of the X-ray emitted would be $hc/17\,400 = 0{\cdot}071$ nm. This X-ray would be known as the K_α X-ray of molybdenum. If, however, an electron jumped from the M shell to fill the vacancy in the K shell the energy difference ΔE would be 19 600 eV (20 000–400 in fig. 2.1 (*b*)) and the X-ray wavelength would be 0·063 nm. This X-ray would be known as the K_{β_1} X-ray of molybdenum. The energies we have quoted for molybdenum are not the same for other elements and the K_α X-ray of each element will have a different energy and wavelength. These

X-rays are known as characteristic X-rays and if we can measure the energy (or the wavelength) we can tell which element must have emitted them. This forms the basis of electron probe microanalysis which we shall meet in Chapter 5.

As well as the discrete electron transitions which give rise to the characteristic K, L and M X-rays many other processes cause energy to be emitted as X-rays when a specimen is bombarded with electrons. However, these other processes do not lead to any particular wavelength of X-ray occurring much more frequently than any other and we speak of this inevitable X-ray background as *white radiation* or Bremsstrahlung (German: ' braking radiation '). However, no X-ray can be emitted with a greater energy than that of the incoming electron. In consequence the total spectrum of X-rays emitted from, for example, molybdenum bombarded with 30 kV electrons would be something like fig. 2.8.

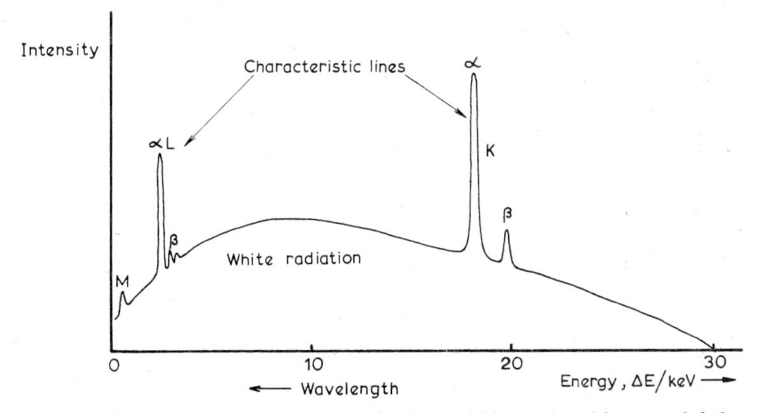

Fig. 2.8. The total X-ray spectrum which would be emitted by a molybdenum specimen which was bombarded with 30 kV electrons. The height of the white radiation has been exaggerated compared with the characteristic peaks in order to illustrate the effects clearly; in reality the K_α peak would be several thousand times more intense than the white background.

This gives us a picture of the X-rays which are generated in the specimen by the electron beam, but of far greater practical importance is which of these get out of the specimen and can be detected. We mentioned in Section 2.5 that X-rays are absorbed by the specimen—the greater their energy (i.e. the shorter their wavelength) the farther they will travel on average. For the majority of materials, K X-rays will be affected only slightly by travelling through a few micrometres of a solid. Since the electron beam is unlikely to penetrate a specimen to a depth of more than a few micrometres we will not be far wrong if we assume for the moment that all the X-rays generated by the electron beam can escape to the surface of the specimen.

Apart from secondary electrons and X-rays there are other secondary effects which we ought to mention, although there is not the space in this book to do justice to their applications in electron microscopy. Thus, just as there are characteristic X-rays whose energy tells the nature of the atom from which they were emitted, there are also characteristic electrons emitted which carry similar information, although detecting them is slightly more difficult. These are called Auger electrons, after their discoverer. Another effect is the emission of light, known as cathodoluminescence. This occurs scarcely at all from some materials but is extremely strong from others. We meet the effect every day when we watch a cathode-ray tube or a television set and in the transmission electron microscope it is the mechanism by which we see any image at all!

Finally we must put in perspective all the various secondary effects. None of the processes which lead to radiation leaving the specimen (as electrons, X-rays or light) is very efficient. In fact, of all the energy carried by the electrons incident on a solid specimen, well over 90% finally has the effect of heating the specimen. Only about 2% of the incident electron energy is converted into X-rays!

2.7. *Diffraction of electrons*

Most of the secondary effects described in the previous section arise whatever the nature or form of the specimen being bombarded with electrons. However, one effect occurs only from a particular type of specimen in certain specific conditions. This is electron diffraction and it is so important to the development and application of electron microscopy that it requires a section to itself.

The diffraction of electrons only occurs when the atoms of the specimen are arranged in some regular way and the elastically scattered electrons within a specimen then travel predominantly in certain directions instead of being randomly scattered in all directions. The most straightforward case to consider is that of a perfect crystal specimen, in which all the atoms lie on very regular sites. Let us consider a simple example in which all the atoms of our material lie on a lattice based on a simple cube. A cross-section of a very thin specimen of this material will then look like fig. 2.9 (*a*). Now if an electron beam is incident on this specimen it will be elastically scattered by some of the constituent atoms; two typical events are shown in fig. 2.9 (*b*). We can think of the electron beam as a wave motion and apply a similar argument to that used to explain the diffraction of light or X-rays. The incident electron beam is

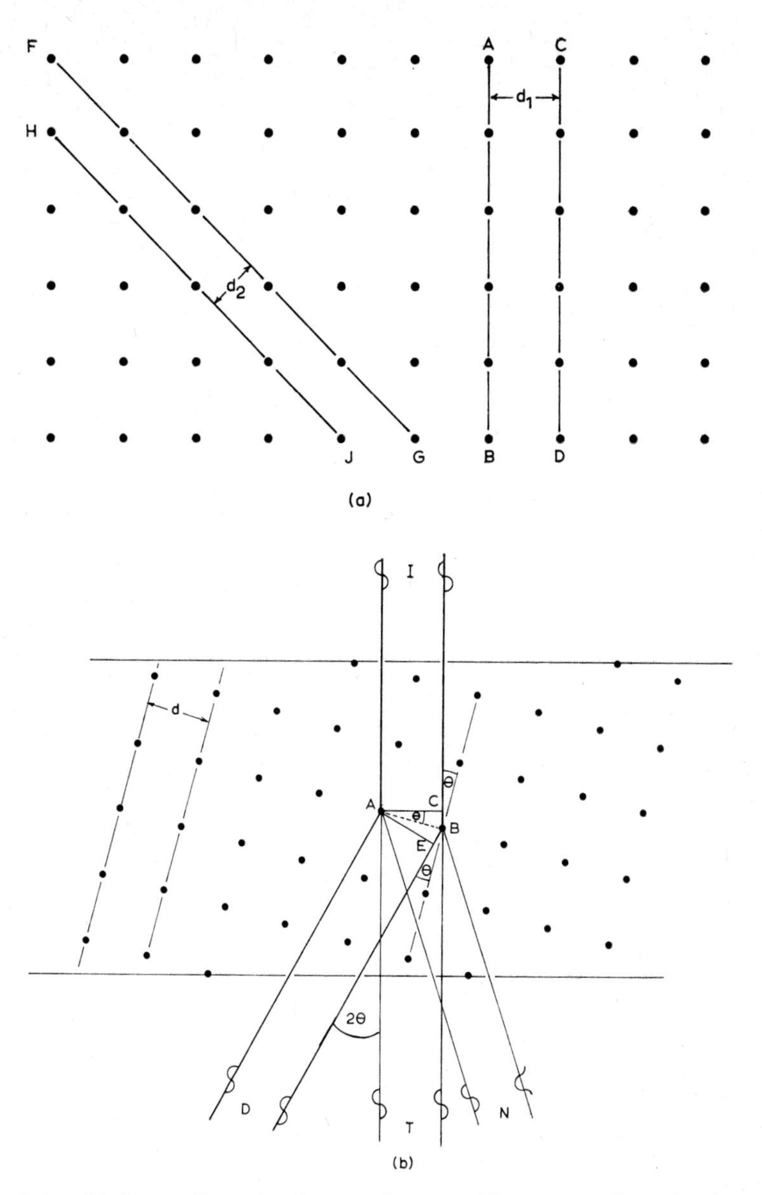

Fig. 2.9. (*a*) A two-dimensional array of atoms with two sets of atomic planes, of spacing d_1 and d_2, indicated. (*b*) The scattering of an incident beam of electrons (I) by a crystalline specimen. Intense beams of electrons may emerge from the other side of the specimen undeviated (T) or having been diffracted from the atomic planes of spacing d (D). In other directions (e.g. N) no intense beams will be formed.

coherent; in other words, all the individual electron waves are in phase, as shown diagrammatically in the upper part of fig. 2.9 (*b*). Any scattered electron waves which are also in phase with one another will reinforce and lead to a strong beam of electrons, whereas any scattered waves which are out of phase will not reinforce. The scattered waves in the direction SD will be in phase if the path length from I to D is the same for both scattered waves, or if the path lengths only differ by an integral number of wavelengths. The only way in which the path lengths can be exactly the same is for the electron waves not to be scattered but to continue as the transmitted wave IST. However, the scattered wave SD can be in phase if the distance CBE is an integral number of wavelengths, or algebraically if $CB + BE = n\lambda$, where n is an integer. However, from the simple geometry of fig. 2.9 (*b*),

$$CB = BE = d \sin\theta$$

and hence

$$2d \sin\theta = n\lambda \qquad (2.8)$$

is the condition for reinforcement. This is well known as *Bragg's Law* and it is applied widely to the reflection and diffraction of light and X-rays as well as electrons. It tells us that we can expect very few elastically scattered electrons to emerge from our specimen unless they are at an angle θ which is a solution of equation (2.8). Since n can be any integer there will obviously be a series of values of θ but the electrons scattered in any other direction, such as SN in fig. 2.9 (*b*), will not be in phase and therefore will not form a significant beam of electrons. The overall consequence is that if a parallel beam of electrons IS in fig. 2.10 is incident on a thin crystalline specimen, several beams of electrons are likely to emerge from the other side. In practice those beams with $n = 0$ and $n = 1$ are much stronger than those of higher order ($n > 1$) and so we often refer to Bragg's Law in the simple form

$$\lambda = 2d \sin\theta. \qquad (2.9)$$

Now we need to consider the significance of d and θ. We have called d the spacing of the atoms which are doing the scattering of the electrons. If the atoms form a crystalline array, as we have assumed in fig. 2.9 (*a*), d is really the spacing of the lines of atoms; for example, d_1 is the spacing of AB and CD. In a real three-dimensional crystal these lines would be planes and so we speak of d as the interplanar spacing. In any crystal there will be other planes we can define, such as FG and HJ, which will have a different spacing, d_2, associated with them. A whole subject, crystallography, has grown up to describe crystals in terms of the positions of their atoms and many elegant ways of describing the planes of atoms have been devised. For our purpose in this elementary text we shall merely note that all crystalline materials can be described in terms of their interplanar spacings and that these have been measured and are tabulated for a wide range of materials. In particular all the metallic elements and many of their alloys exist in crystalline forms and are well documented.

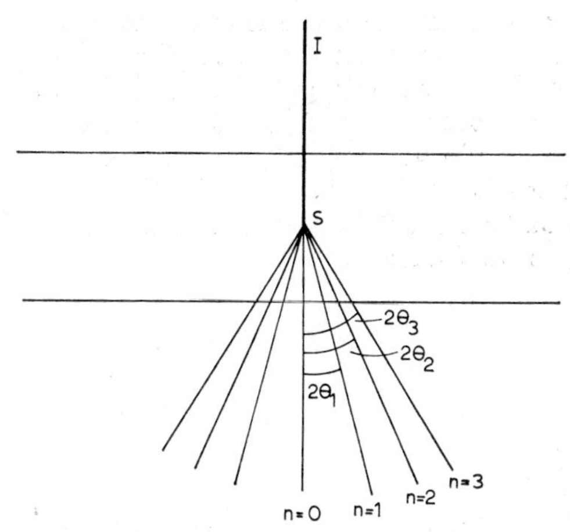

Fig. 2.10. The diffraction of an incident electron beam IS by a crystalline specimen. As well as the undeviated beam ($n=0$) the first, second and third order diffracted beams are shown, corresponding to $n=1$, $n=2$ and $n=3$ in Bragg's Law.

Now let us consider the significance of the angle θ, the 'Bragg angle'. Because of the very short wavelength of the electrons used in electron microscopy we can simplify equation (2.9) even further. If we substitute typical values of θ and d in equation (2.9), say $\lambda=0{\cdot}0037$ nm (the value for 100 kV electrons from table 2.1) and $d=0{\cdot}4$ nm (the spacing of some commonly occurring planes in crystals of aluminium), we find that $\sin\theta=0{\cdot}0046$ and hence that $\theta\approx0{\cdot}0046$ rad, since for such small angles we can approximate $\sin\theta$ to θ without an appreciable error. In fact this 'small angle approximation' is valid for angles up to at least 5° and since we almost never encounter angles greater than this in electron diffraction we can replace $\sin\theta$ in equation (2.9) by θ and write

$$\lambda=2d\theta. \qquad (2.10)$$

Since θ is always so small it is obvious that we have exaggerated the angles in figs. 2.9 and 2.10 and that in reality a vertical electron beam will only be strongly diffracted from planes of atoms such as AB in fig. 2.9 (*a*) which are almost vertical themselves.

To what use can we put diffraction? First let us redraw fig. 1.2 to include a few more ray paths so that we can see what happens in a microscope to rays which are parallel when they leave the specimen. Figure 2.11 shows just two sets of parallel rays leaving the specimen AA′. Before forming the first image at BB′ all rays parallel to the two single-arrowed rays will pass through the point D_1 and all rays parallel to the double-arrowed rays will pass through D. For every set of parallel rays leaving the specimen there is a corresponding point in the plane DD_1 (known as the back focal plane of the objective lens). Similarly, if we follow the ray paths through the second (projector) lens we find that a second set of points is formed in the plane EE_1. If we were to put a viewing screen or a photographic plate at EE_1 instead of at CC′ we would see a *diffraction pattern* instead of the image. This pattern contains a wealth of useful information about the specimen, as we shall see.

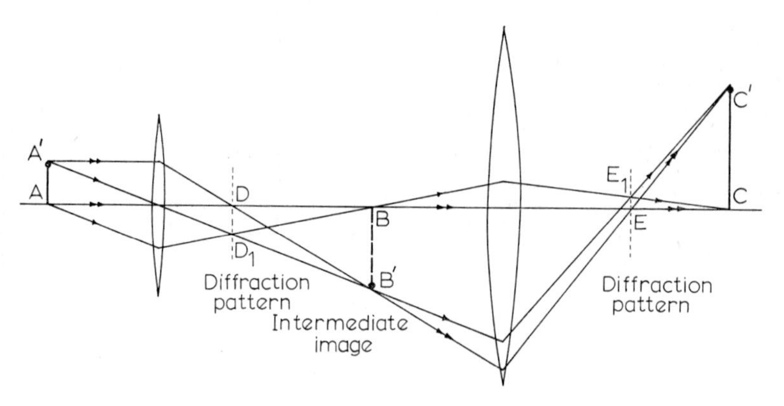

Fig. 2.11. The ray diagram for a two-stage projection microscope (see fig. 1.2) showing the formation of diffraction patterns at DD_1 and EE_1.

Let us now go back and consider what happens to the diffracted beams of electrons from our thin crystalline specimen. Each beam will be brought to a point and these points will form a diffraction pattern in the electron microscope. Figure 2.12 (*a*) is a very simplified diagram which shows schematically the formation of two spots in the diffraction pattern from the undiffracted beam and one first order ($n=1$) diffracted beam at an angle 2θ from it. In order to make calculations about the crystal structure of the specimen from the size and shape of the diffraction pattern it is easier to consider the electron microscope as if it consisted of no magnifying (projector) lenses after the specimen, as shown in fig. 2.12 (*b*). The diffraction pattern would now be farther from the specimen, at a distance known as the *camera length*, *L*, but the overall geometry is exactly equivalent and particularly the distance between the two spots, r, is identical. It is then easy to see that

$$\frac{r}{L} = \tan 2\theta$$

and, using the small angle approximation again to say $\tan 2\theta = 2\theta$, we can write

$$\frac{r}{L} = 2\theta. \tag{2.11}$$

Since from equation (2.10) we have that $2\theta = \lambda/d$ we can combine the two equations to find

$$\frac{r}{L} = \frac{\lambda}{d} \quad \text{or} \quad r = \frac{\lambda L}{d}. \tag{2.12}$$

Since L is only a notional distance, the configuration in a real electron microscope being nearer that of fig. 2.12 (*a*), we often combine the two constants λ and L and call λL the *camera constant*. This makes it very plain that the distance from a spot to the undiffracted spot, r, in an electron diffraction pattern is inversely proportional to the spacing

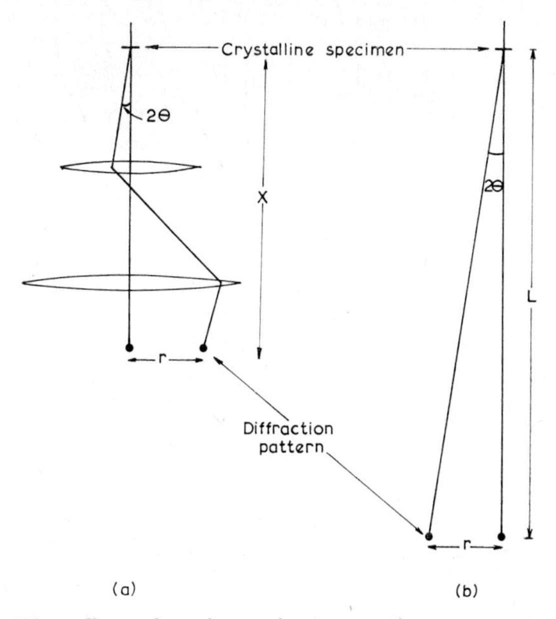

Fig. 2.12. The effect of projector lenses on the geometry of diffraction patterns. The action of the two lenses means that the actual distance between the specimen and the diffraction pattern, X, is not the same as the effective ‘ camera length ’ L.

of the crystal planes which gave rise to that spot. We therefore have a way of measuring interplanar spacings for crystalline materials; in fact there is much more information which it is possible to deduce from an electron diffraction pattern such as that shown in fig. 2.13, as will appear in the next two chapters.

Fig. 2.13. A diffraction pattern from a single crystal of aluminium.

The phenomenon of diffraction is not confined to perfectly crystalline materials. In general diffraction will occur from any feature in a specimen which scatters electrons and is fairly regularly spaced. There are many such features to be found in the microstructures of both biological and non-biological materials, and some of them will be mentioned later in this book. However, the major significant feature of all diffraction is the inverse relationship between the spacings of the features in the specimen and the spacing of the corresponding features in the diffraction pattern. To use the example of a crystalline metal lattice which we have already been dealing with, we would find that the diffraction pattern from copper contained the same pattern of spots as the pattern from aluminium, because these two metals have the same crystal structure. However, the spacing of the spots from copper would be *larger* than that of the aluminium spots since the main lattice spacing of copper, 0·36 nm, is *smaller* than the 0·4 nm of aluminium.

CHAPTER 3

the transmission electron microscope

3.1. *How it works*

By the end of the 1920's all the components necessary to construct a microscope using electrons had been developed. We have seen in the last chapter how the illumination can be provided by an electron gun and how lenses can be made using fairly simple electromagnetic fields. The remaining essential components are a viewing screen—easily provided in the form of a layer of electron-fluorescent material such as zinc sulphide—and a camera. Fortunately most photographic emulsions are as sensitive to electrons as they are to light and new materials did not have to be developed. In the early 1930's these components were first put together to form a transmission electron microscope (TEM) exactly analogous to the transmission light microscopes in everyday use by biologists. The most obvious, but trivial, difference is that for convenience the illumination (electron gun) is generally put at the top of the microscope and the viewing screen at the bottom; this is the exact inverse of the arrangement of most light microscopes. This inverted arrangement has persisted for 40 years; not until 1971 was a commercial electron microscope available with the viewing screen at the top and the electron gun on the floor!

The early microscopes generally used one lens as a collimator to control the electron beam before it reached the specimen and one or two projector lenses to magnify the subsequent image. They were therefore exact electron-optical analogues of fig. 1.3 (*a*). Nowadays, however, it is more common to find two condenser lenses and three or four projector lenses as indicated in fig. 3.1 (*a*). The electronic circuitry necessary to control the electron gun, five or six lenses and a vacuum system is quite complicated, and the actual microscope which corresponds to the section shown in fig. 3.1 (*a*) is therefore apparently a mass of knobs and dials, as fig. 3.1 (*b*) shows. However, once the optical principles of microscopy are understood even the most sophisticated electron microscope is very simple to operate.

Let us take the modern six-lens electron microscope shown in fig. 3.1 as an example and examine it in a little more detail. At the top of the instrument, two or three feet above the seated operator's head, is the electron gun. The most common types of TEM have

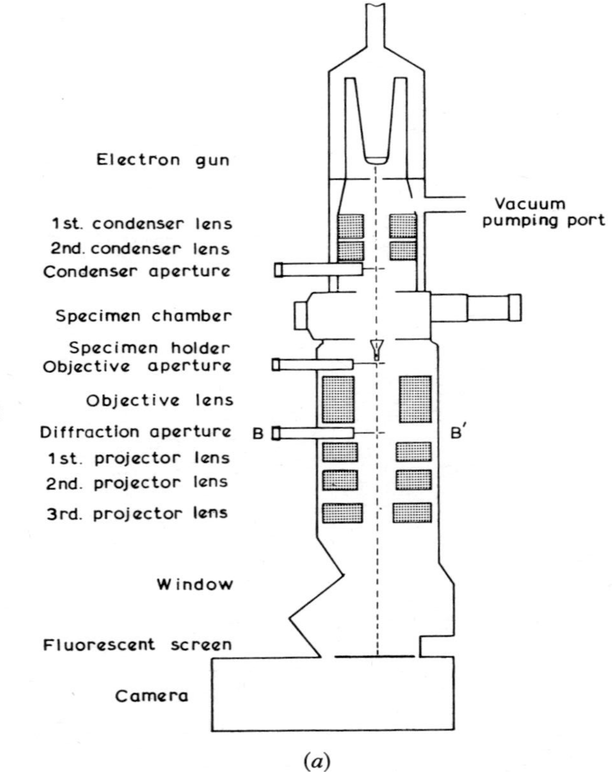

(*a*)

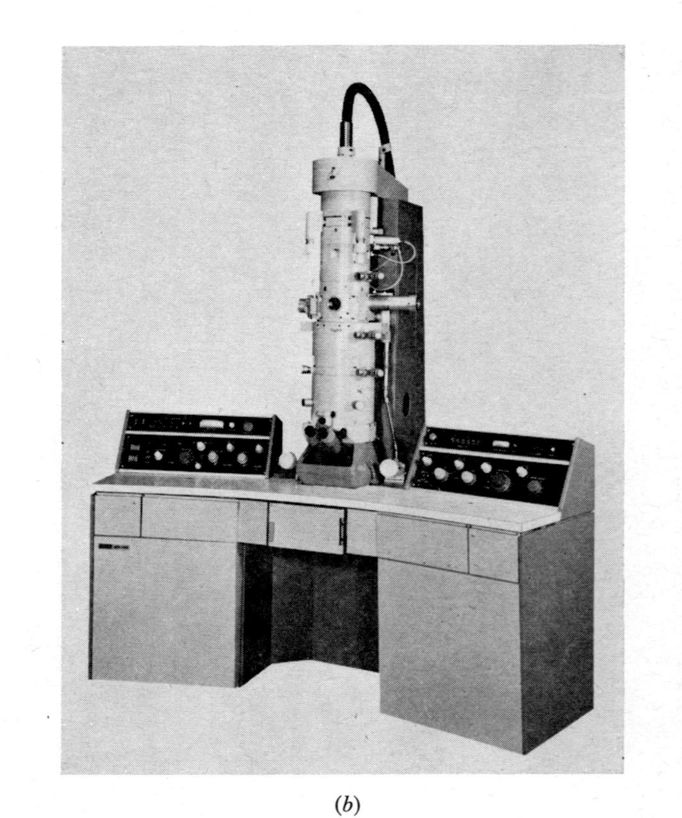

(*b*)

Fig. 3.1. (*a*) A schematic cross-section of a modern six-lens transmission electron microscope. (*b*) A photograph of the actual microscope which corresponds to this cross-section (JEOL JEM 100B).

guns capable of accelerating the electrons to voltages between 20 kV and 100 or 125 kV. The appropriate voltage to use will depend on the nature of the specimen. For some applications, particularly if the specimen is very thick, it is an advantage to use much higher accelerating voltages. Microscopes capable of 1 MV (one million volts) or more are available but they are extremely large and expensive, costing in the region of a quarter of a million pounds at the time of writing, and only a few are in use.

Below the electron gun are two condenser lenses whose main function is to control the diameter of the beam of electrons as it hits the specimen. This is necessary since the area of the specimen which must be illuminated depends on the magnification at which it is to be viewed. Since we often want the final image to be as bright as possible it is not sensible to ‘ waste ’ electrons by illuminating an area say 1 mm in diameter when we are only viewing an area which is 10 μm across. We therefore condense the illuminating beam until it is only a little larger than our field of view.

Below the condenser lenses lies the specimen chamber, which is one of the most crucial parts of the microscope, since so many delicate mechanisms must be built into a small space. The essential requirement is that the specimen, which is likely to be about 1 mm square and so thin that it may be hardly visible, must be held in exactly the correct position in relation to the objective lens. Since it is not very useful to be able to study only one area of the specimen we need a mechanism which will move the specimen in a horizontal plane to enable us to look around the specimen and select the most interesting parts of it to photograph. Not only must this mechanism be so fine that it can move the specimen smoothly over distances as small as 1 μm, but when the appropriate area for high magnification photography has been selected the specimen must not move as much as 1 nm per second (1 mm in 12 days!) or the resulting micrograph, taken with a photographic exposure of several seconds, will be blurred. On top of this very stringent condition it is extremely useful if the specimen can be tilted at an angle of up to 45° so that the electron beam can be made to strike it at a great variety of angles. Add to this the requirement that we must be able to remove the specimen from the microscope and replace it with another one in a matter of minutes (or preferably seconds) and you begin to realize how sophisticated this part of the microscope is. Three specimen holders, one of which enables the specimen to be tilted, are shown in fig. 3.2 with the specimen on its copper grid lying beside them. To enable the changing of specimens to be carried out in a very short time the specimen chamber usually contains an airlock so that only a small volume needs to be evacuated by the vacuum pumps each time a specimen is introduced, instead of the whole microscope.

Immediately below the specimen is the objective lens, in the middle of which runs an aperture holder which carries three or four circular pinholes of different sizes. These apertures are exactly analogous to the objective aperture shown in the light microscope ray diagram in fig. 1.3 (*a*) and are situated in the back focal plane of the objective lens. In the electron microscope they need to be very small, in order to keep the angular aperture α (shown in fig. 1.9) smaller than 10^{-2} rad ($\sim\frac{1}{2}°$). Normally holes of diameter 20 μm to 100 μm in platinum or molybdenum sheets are used. The choice of aperture defines the numerical aperture (NA) of the objective lens and hence α and the ultimate resolution of the microscope (see equations (1.2) and (1.4)). Normally a compromise decision is reached—a larger aperture will make the resolution poorer because spherical aberration becomes important, but a smaller aperture also limits the resolution because of the diffraction effect discussed in Chapter 1 (Section 1.4).

Fig. 3.2. Three specimen holders for transmission electron microscopes. The holder on the far right enables the specimen to be tilted through 60° about two perpendicular axes. Between the holders are two copper specimen support grids, which are 3 mm in diameter.

The objective aperture has another role, as we shall see later in this chapter, in that it improves the *contrast* in the image. Without an aperture in place, the image of almost any specimen appears uniformly grey, rather than black and white. The reason for this is discussed in the next section.

An intermediate image is formed by the objective lens at a magnification of about 100 × . By fine control of the current passing through the lens this image is focused on the plane BB′ in fig. 3.1. In this plane is another set of apertures by which different sized areas of the field of view can be selected for such purposes as forming electron diffraction patterns. These apertures are very similar to the objective apertures but are known as *field limiting* or *diffraction* apertures. The intermediate image is enlarged to the final magnification required (which may be as high as 1 000 000 × in some modern microscopes) by two or three *projector lenses*.

The image is viewed on a fluorescent screen lying a few tens of centimetres below the final projector lens. The operator can see the image through a glass window in the vacuum system either with his unaided eye or using a pair of binoculars to help him focus fine detail. When a photographic record is required a hinged part of the fluorescent screen is lifted up and the electrons are allowed to fall on to a photographic plate or film in the camera below. Normally exposures of between 1 s and 10 s are used and since the photographic emulsion is sensitive to light as well as electrons the microscope room must be completely dark while the exposure is made. Since it is necessary to break the vacuum system to remove the plates most electron microscope cameras hold 20 or 24 plates and this number of exposures can be made before it becomes necessary to stop operating and let air into the microscope while the plates are changed. Although the camera may be several centimetres below the fluorescent screen, the image recorded on the plate is in just as sharp focus as the image which is viewed on the screen: this makes life very much easier for the microscopist and is due to the enormous *depth of focus* of an optical system when its numerical aperture is very small. At high magnifications the image is in sharp focus for hundreds of metres below the fluorescent screen, incredible though this may seem.

Our discussion has now reached the very bottom of the microscope shown in fig. 3.1 but we have not accounted for more than half the controls visible in fig. 3.1 (*b*). Most of the remaining electronics is needed to correct the various aberrations which have been discussed in Chapter 1. If spherical aberration is to be minimized it is imperative that the electron beam travels exactly down the optical axis of the microscope. Several sets of coils, and their associated controls, are necessary to align the electron beam within the microscope and ensure that this is so. Similarly the astigmatism which arises because of contamination of the very small apertures can be corrected by further sets of coils. For each of these coils there must be a corresponding set of controls on the microscope desk, but fortunately for the operator most of them need adjusting only rarely. The four essential sets of controls are these: the condenser lens current which determines the area of specimen illuminated; the objective lens current which focuses the image; the projector lens current which controls the magnification; and the specimen shift controls which enable the operator to look around his specimen.

3.2. *Why we see anything in the image*

So far we have discussed both the transmission electron microscope and the light microscope solely in terms of *geometrical optics*. In other words we have constructed ray diagrams to show that if there is an *object* (the specimen) at a certain point, then there should be a corresponding *image* at another point. We choose to put our fluorescent screen or our photographic camera at the image point but geometrical optics does not tell us anything about the appearance of the image from any particular specimen. It would be very easy, as was mentioned in the previous section, to form an image in the electron microscope which was almost uniformly grey. Such an image might be perfectly in focus and at just the right magnification but if it does not show any *contrast* between dark areas and light areas then we can learn nothing from it. We must therefore consider what contrast mechanisms might operate when a thin specimen is viewed in a TEM. The major mechanisms are scattering contrast, which is of prime concern when looking at biological specimens, and diffraction contrast, which is employed when crystalline specimens are viewed. Figures 3.3 and 3.4 respectively are electron micrographs of typical biological and crystalline materials—we shall use them to illustrate the two types of contrast mechanism.

Scattering contrast will be dealt with first, for three reasons: two-thirds of all electron microscopists are biologists; it is easier to understand; and it has strong similarities to the light microscope situation. Let us first consider why certain areas of a specimen appear dark in a light microscope. The fundamental (and obvious) point is that very little light reaches the image from the ' dark ' parts of the specimen. What is less obvious is that there are two possible reasons for this: either the light has not managed to penetrate the ' dark ' region of the specimen, in other words it has been absorbed, or some light has been transmitted but it is scattered through such an angle that it does not get through the objective aperture into the objective lens and hence to the image, as shown for example in fig. 1.3 (*a*). In the light microscope the objective aperture is usually very large so that light must be scattered through almost 90° before it fails to pass through the aperture, consequently the scattering of light does not contribute much to the contrast of the image, which is determined almost wholly by the absorption of light in the different areas of the specimen. However, in the electron microscope the apertures are

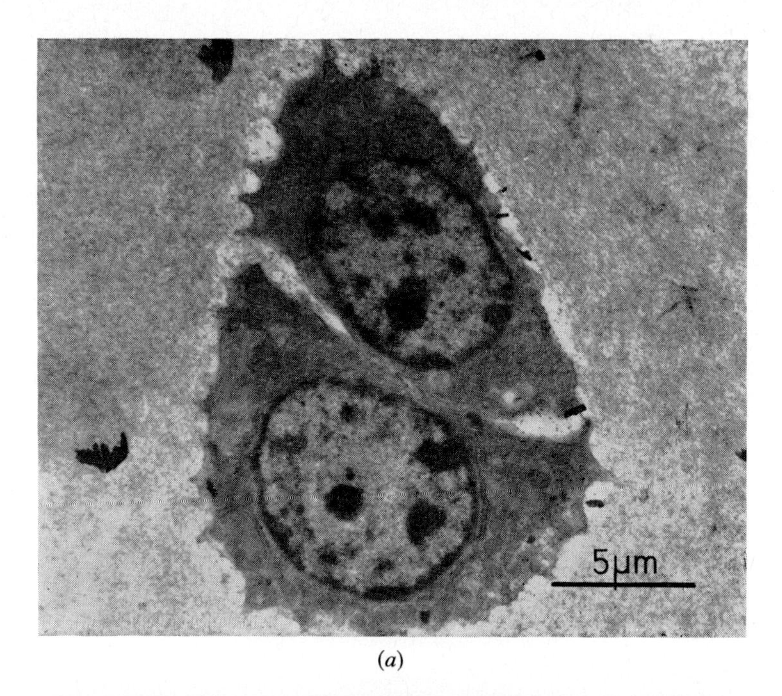

(*a*)

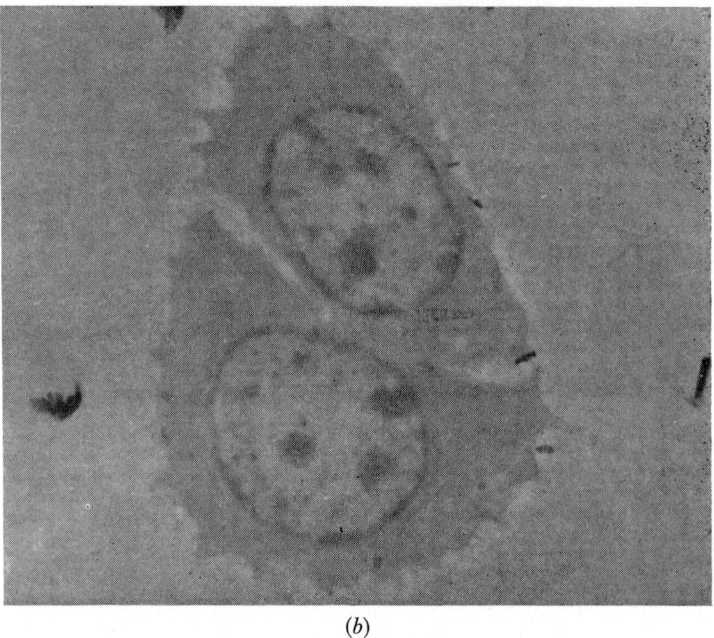

(*b*)

Fig. 3.3. An animal cell (a cartilage cell from chick trachea) photographed (*a*) with and (*b*) without an objective aperture in position.

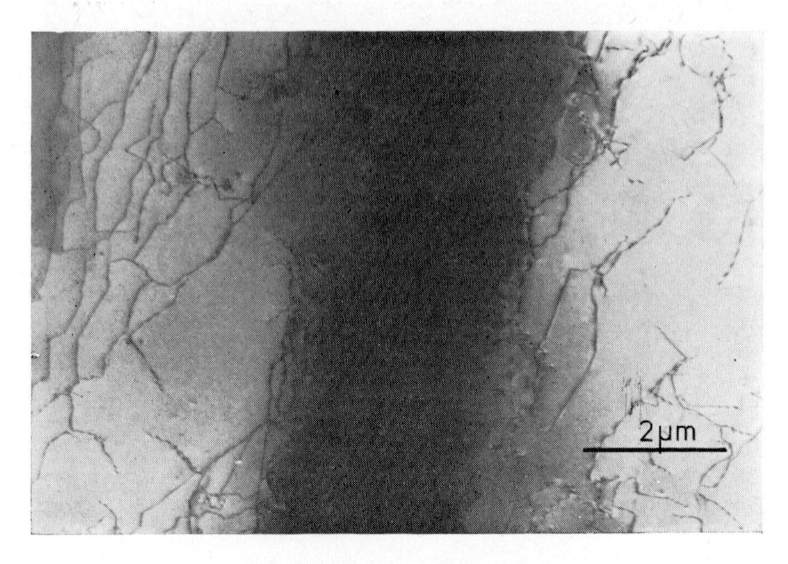

Fig. 3.4. A micrograph showing a prominent extinction contour and several dislocations in a thin foil of aluminium.

very small, as indicated in fig. 3.5, and electrons can quite often be scattered through a large enough angle to be stopped. In order to avoid chromatic aberration (fig. 1.10) it is necessary to use very thin specimens for electron microscopy so that only a very small percentage of electrons are truly absorbed by the specimen. In this case scattering plays a much bigger part and is the major contributor to contrast in the final image.

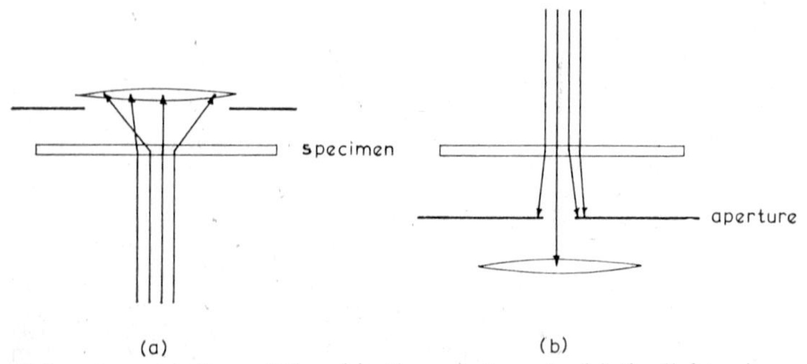

Fig. 3.5. A comparison of the objective apertures in (*a*) the light microscope and (*b*) the electron microscope. The aperture is always extremely small in the electron microscope and hence electrons which are scattered only a small angle from their initial direction are prevented from contributing to the image.

The amount of electron scattering which occurs (i.e. the percentage of electrons passing through a certain region of the specimen which are scattered) depends on the number of atoms in that region of the specimen and their mass. Thus a thick part of a specimen will scatter more than a thinner part and a region containing heavy elements will scatter more than a region of the same thickness containing lighter elements. These effects are illustrated in fig. 3.6, from which it can be seen that at any point below the specimen the intensity of the transmitted beams of electrons (i.e. those electrons which have not been scattered) depends on the thickness and the mass of the specimen at that point. The total number of electrons passing through each part of the specimen is very nearly the same, but the proportion of scattered electrons will depend on the structure of the specimen. If no objective aperture is used in the microscope then virtually all the electrons, scattered and unscattered, will pass into the objective lens and on to the image. This is because the scattering is so strongly peaked in the forward direction, as fig. 2.5 showed, that very few electrons are scattered through large angles. If a small objective aperture is inserted, however, then the scattered electrons will be stopped and only the unscattered electrons will contribute to the image. The striking difference which the insertion

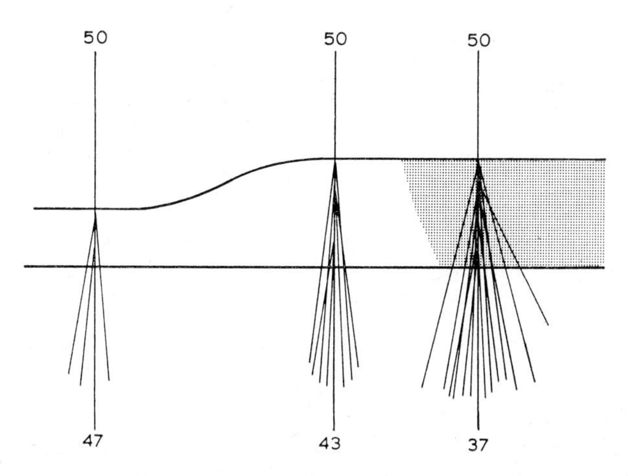

Fig. 3.6. The scattering of electrons from different regions of a thin specimen. In a thin area (on the left) only a few electrons are scattered and perhaps 47 of the original 50 incident electrons continue undeviated. In a thicker region of the same material (centre) more are scattered and perhaps only 43 remain in the undeviated beam. From a region of the same thickness but which contains heavier atoms even more scattering will take place and perhaps only 37 electrons continue to pass through the objective aperture.

of an aperture makes is illustrated in fig. 3.3 (*a*) and (*b*). The smaller the aperture the better the contrast seen in the image but, as discussed in Chapter 1, there is a fundamental limitation on the resolution attainable because of diffraction in the aperture and therefore it is not possible to use very small apertures if high resolution is needed.

We are now in a position to understand the micrograph of an animal cell shown in fig. 3.3. The section should be of a uniform thickness and therefore we do not expect any contrast to arise from local differences in specimen thickness. However, biological tissue consists for the most part of carbon, oxygen, nitrogen and hydrogen atoms and no part of the specimen is very much more dense than any other part. In order to produce a reasonable amount of contrast in the image it is therefore necessary to artificially ' stain ' selected features of the specimen so that they scatter electrons effectively. The procedures necessary are similar to those used for optical microscope specimens and will be explained in more detail in a later section. For the moment we need to think of a staining procedure as attaching atoms appreciably heavier than carbon or oxygen to known parts of the specimen. In the cartilage cell shown in fig. 3.3 uranyl acetate has been used as a stain, since it is known to attach itself to fatty regions of the specimen. Consequently we see the nucleolus and the nuclear membrane, which contain a great deal of lipid, as dark areas in the micrograph. To improve the contrast of other features other stains are needed, so clearly the microscopist has to know what his stain will show up before he can interpret his micrographs.

Fortunately the micrographs produced from stained biological specimens bear a very great similarity to light micrographs of the same material, so that interpretation of the contrast in the image is normally not difficult. However, the second major type of image contrast, diffraction contrast from crystalline specimens, is unlike any effect normally seen in a light microscope and needs more subtle interpretation. The materials scientist, who is frequently dealing with crystalline materials, needs to have a clear understanding of electron diffraction before he can interpret many of his micrographs. We only have the space here to outline the basic principles of diffraction contrast and to illustrate a few of the features which can be made visible in a crystalline specimen.

The simplest possible crystalline specimen consists of a perfect single crystal of identical atoms. Since at all points in the specimen the density is the same and the species of atom is the same we can expect to see no contrast in the image between one part and another. Scattering contrast, in other words, is of no importance. However, the actual brightness of the whole image depends very critically on the angle between the electron beam and the specimen, so that it is possible to get a completely ' black ' image or a completely ' white ' image from the same crystal specimen just by tilting it. In order to understand how this arises we must think about the possibility of diffraction from the specimen, in terms of Bragg's Law (equation (2.9)). Figure 3.7 (*a*) shows a perfectly flat crystalline specimen aligned so that the electron beam is exactly at the Bragg angle to a certain set of lattice planes (lines of atoms). Since Bragg's Law is obeyed, diffraction will occur and the majority of electrons will leave the specimen at an angle 2θ to the incident beam. If an objective aperture is in position these electrons will be stopped and virtually no electrons will be able to contribute to the image, which will therefore appear dark. On the other hand, if the specimen is tilted a small amount away from the Bragg condition (i.e. θ is changed by as little as $\frac{1}{4}°$, exaggerated in fig. 3.7 (*b*)), no diffraction will take place from these lattice planes and most of the incident beam will pass through the objective aperture. The image will therefore appear uniformly bright.

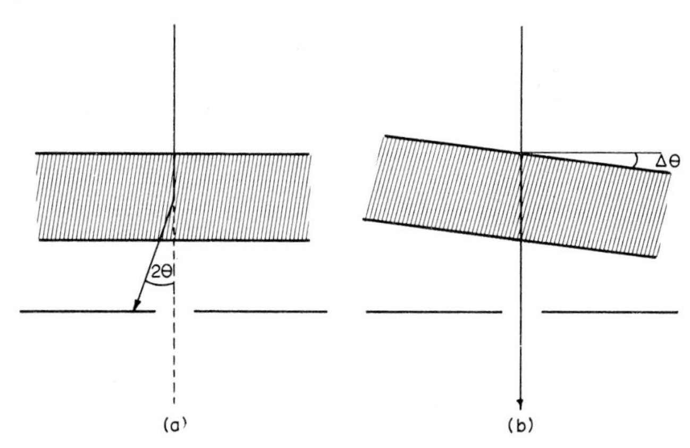

Fig. 3.7. (*a*) A crystalline thin specimen aligned so that the electron beam strikes one set of lattice planes exactly at the Bragg angle. The majority of electrons in the beam are diffracted and few continue undeviated and pass through the objective aperture. (*b*) The same specimen tilted a small angle $\Delta\theta$ from its previous position. Even if $\Delta\theta$ is as small as $\frac{1}{4}°$ the lattice planes will be so far from the Bragg angle that a negligible amount of diffraction will occur and virtually the whole of the electron beam passes through the aperture.

Most real specimens are not perfectly flat, however, but because they are so thin they become slightly buckled. The lattice planes, therefore, are not all at the same angle to the incident beam; certain parts of the specimen will be at the Bragg angle and will appear dark, while other areas will lie away from the Bragg angle and will appear bright. This effect is shown schematically in fig. 3.8. It leads to the appearance of dark lines, called *extinction contours*, even in the image of a piece of perfect crystal. Several extinction contours in a specimen of aluminium are shown in fig. 3.9; they are not a very useful part of the image but it is extremely important that we know what they are, since almost every image of a crystalline specimen will contain some. It is of far greater interest to the materials scientist or the physicist to know what imperfections are present in a crystalline specimen and fortunately diffraction contrast is an excellent way of making visible the majority of defects which do occur in crystalline materials.

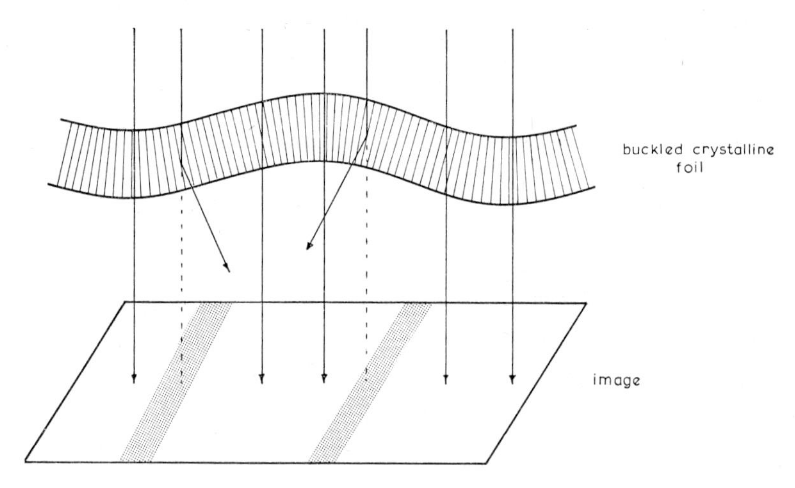

Fig. 3.8. A buckled single-crystal foil showing the formation of extinction contours in the image corresponding to regions of the foil which were exactly at a Bragg orientation.

Most metals and alloys are used in their polycrystalline form, in which many small regions of fairly perfect crystal (grains) are present in different orientations in the solid. This grain structure is usually examined with a light microscope operating in reflection (fig. 1.3 (*b*)), using a polished face of the metal as a specimen. A typical light micrograph, showing the grain boundaries and some 'twin' defects within the grains, is shown in fig. 3.10. For examination in the electron microscope only a very thin section of the specimen can be used and micrographs are usually taken at such high magnifications (several thousand times instead of the few hundred of fig. 3.10) that only one of the grains is studied at a time. Therefore the part of the specimen seen in the electron microscope is usually effectively a single

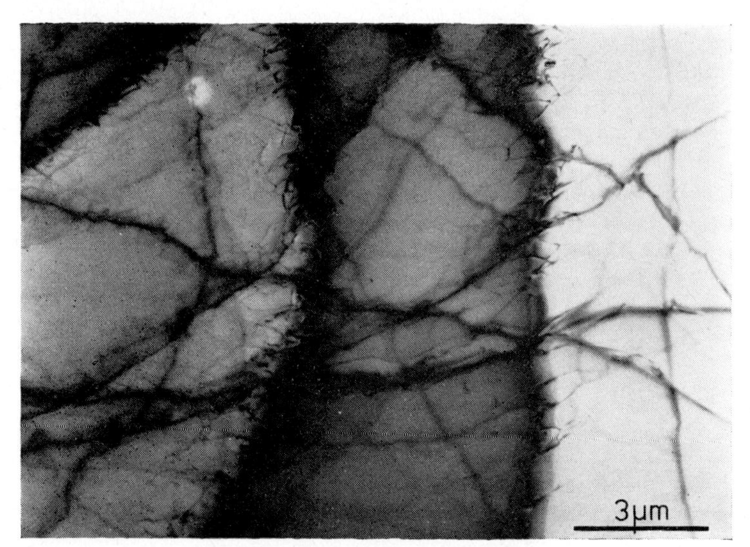

Fig. 3.9. A buckled region of an aluminium foil which shows many extinction contours. Several dislocation tangles are visible at the edge of the major contours.

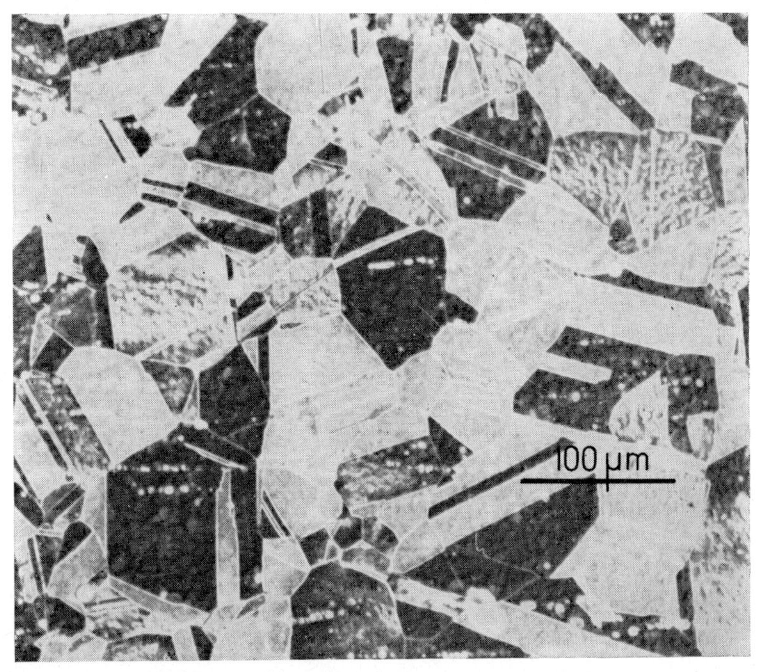

Fig. 3.10. A light micrograph of a polished and etched brass specimen.

crystal, although occasionally a grain boundary or a twin boundary is seen. Even a single grain is never really a perfect crystal, however, and always contains small defects of one sort or another. All the possible defects affect the image in one of three ways; either they bend the lattice planes locally, or they cause interference between the diffracted electron beams from two nearly perfect overlapping crystals, or they alter the local thickness of the specimen. Let us consider each of these types of defect in turn.

Crystal defects which cause local bending of lattice planes are extremely common in metals and alloys. Two of the most frequently occurring are the dislocation, which is responsible for most plastic deformation, and the precipitate, which results from the clustering together of atoms of a single type in an alloy. The ways in which these defects bend lattice planes are shown diagrammatically in fig. 3.11. Far from the defect (where ' far ' in these terms may only mean 10 nm) the planes of atoms are scarcely disturbed by the dislocation or precipitate. However, very close to the defect the lattice is bent quite appreciably and it is this change in orientation of the lattice planes which enables the defect to be seen. If a large part of the specimen is in the orientation to the electron beam shown in fig. 3.7 (*b*) (i.e. only a small angle $\Delta\theta$ away from the Bragg angle),

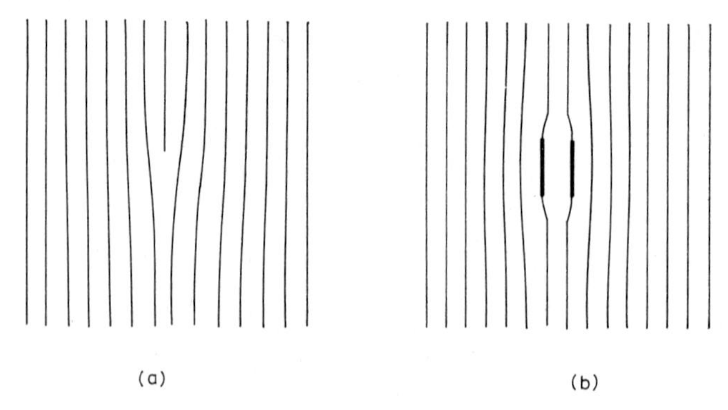

Fig. 3.11. Two types of crystal defect which locally bend the lattice planes. (*a*) The dislocation; (*b*) the coherent precipitate.

then its image appears white. However, a dislocation in this specimen will tilt the lattice in its neighbourhood so that on one side or other of the defect the lattice planes are now at the Bragg angle. From this part of the specimen diffraction will occur strongly and, with an objective aperture in place to stop the diffracted electrons, a dark area will appear in the image which corresponds to the dislocation. The lines visible in the micrograph shown in fig. 3.4 are in fact dislocation lines. Precipitate particles or any other defect which tilts the lattice in its vicinity will give rise to similar effects in the image. It would appear at first sight that we need a rather special set of circumstances to make this type of defect visible, since the specimen needs to be just a small angle from the Bragg angle. However, two factors make this situation very easy to find: there are many sets of lattice planes in any crystal, for each of which there is a Bragg angle—the specimen need only be near one of these orientations. Secondly since the specimen is almost inevitably buckled, extinction contours will be present. These dark lines tell us where the crystalline specimen is exactly at the Bragg condition—we merely need to look either side of an extinction contour to see an area which is near the Bragg condition (fig. 3.4 shows this clearly).

The second type of crystal defect mentioned earlier is that in which two otherwise perfect crystals are overlapped. One obvious case in which this happens is the grain boundary. Although a specimen for transmission electron microscopy is very thin the grain boundaries do not necessarily run exactly perpendicular to the specimen but may slope across it as shown in fig. 3.12 (*a*). When at the grain boundary the electron beam is therefore passing through a crystal of one orientation, giving rise to diffraction, and then through a crystal of a different orientation. Very frequently the diffraction of the electron beam occurring in the two crystals combines to give rise to fringes parallel to the intersection of the boundary and the specimen surface,

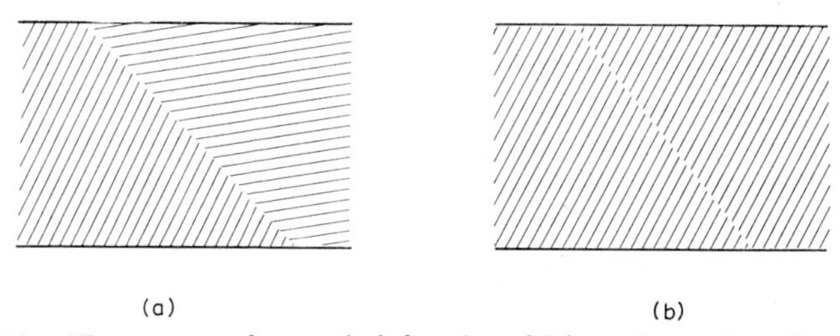

Fig. 3.12. Two types of crystal defect in which regions of perfect crystal overlap. (*a*) A grain boundary; (*b*) a stacking fault.

as shown in fig. 3.13. A similar effect occurs at a stacking fault, which is an area within a single grain where the lattice planes are slightly out of register (fig. 3.12 (*b*)). Although they give rise to similar fringe effects, grain boundaries, stacking faults and twin boundaries can in practice normally be distinguished by their shapes in the image.

A third type of defect which can be made visible when it occurs in a crystalline specimen is one which alters the effective thickness of the specimen. Occasionally specimens may be pitted on a very fine scale and thus truly alter their thickness in certain areas. However, rather more interesting are defects inside the body of the specimen which have a similar effect. There are two common types of defect which fit this description: voids, in other words three-dimensional holes in the crystal lattice, and bubbles which only differ from voids in being filled with gas atoms. Since the gas atoms are few in number and often of rather low mass (for example, helium) the net effect of a void or a bubble is that the specimen contains fewer atoms across its thickness than it would normally; this has the same effect on the scattering of electrons as if the specimen were locally thinner.

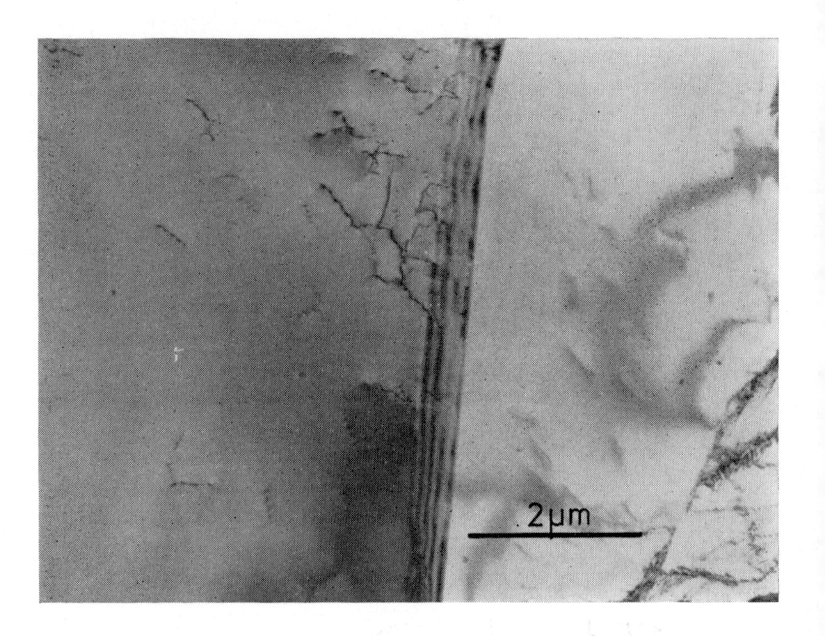

Fig. 3.13. Fringes at the grain boundary between two regions of perfect crystal in a thin foil of aluminium.

Figure 3.14 should make this clear. Both voids and bubbles tend to occur in materials which are subject to neutron bombardment; for example, material used in nuclear reactors. An example of bubbles created in this way is described later in this chapter. Figure 3.33 shows that the bubbles appear as brighter patches against the background of the micrograph but it should be mentioned that, for reasons beyond the scope of this book, it is possible for bubbles and voids to appear darker than the background in some circumstances.

It might seem from the foregoing few pages that the mechanism by which any structure is made visible in the electron microscope is very different for biological and crystalline materials. However, we should remember that identical electron–atom interactions are occurring in both cases, leading to both inelastic and elastic scattering of the electrons. Theoretically therefore all the contrast mechanisms that have been described will operate in all specimens. It just so happens, however, that biological materials rarely contain a large number of periodically stacked molecules and therefore do not give rise to 'diffraction contrast' and also that crystalline materials, particularly metals, rarely contain sizeable regions of very different density and therefore do not give rise to appreciable 'scattering contrast'. An example which illustrates that this is not universally the case can be found in the description of a micrograph of bone later in this chapter. This biological material contains fibres of collagen which consist of highly crystalline material (fig. 3.32).

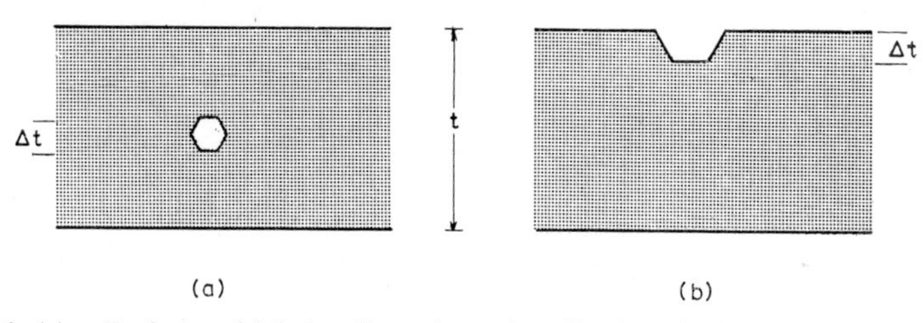

Fig. 3.14. Defects which locally reduce the effective thickness of a specimen (*a*) A void or bubble; (*b*) a surface pit.

3.3. *Electron diffraction and what it can tell us*

The most important thing we can learn from a diffraction pattern in the electron microscope is whether the part of the specimen being examined is crystalline, or amorphous, or contains regions of both types. Crystalline specimens give rise to patterns of spots, as discussed at the end of Chapter 2; fig. 3.15 (*a*) shows diagrammatically the type of diffraction pattern which arises from a single crystal. It consists of spots in a regular array whose spacing from the centre of the pattern is inversely proportional to the distance between the lattice planes. From this pattern we can learn a great deal about the crystal structure of the specimen. If the individual crystals (grains) of the specimen are small, so that, for example, three crystals of the same type but in different orientations are selected by the diffraction aperture, then a more complicated pattern containing three different arrays of spots is seen (fig. 3.15 (*b*)). It is noticeable already that these spots tend to lie on circles and if the crystal size of the specimen is even smaller and many crystals lie within the diffraction aperture the spots do in fact appear to join up and a 'ring pattern' is seen (fig. 3.15 (*c*)). If the crystal size is extremely small it is difficult to define a crystal at all and we refer to the specimen as amorphous, meaning that instead of the majority of atoms being arranged in straight lines and a fixed distance apart the constituent atoms are found in random positions at a variety of distances from one another. Nonetheless, on average the atoms are a certain distance apart and the diffraction pattern from an amorphous material therefore contains a rather diffuse halo, as indicated in fig. 3.15 (*d*).

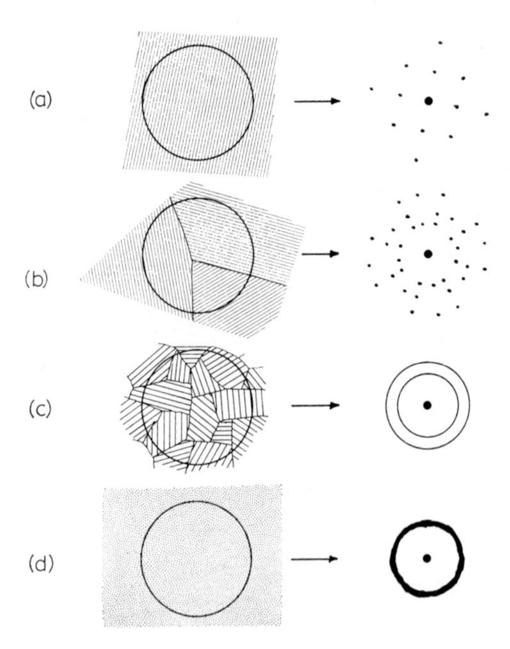

Fig. 3.15. The types of diffraction pattern which arise from different specimen microstructures. (*a*) A single perfect crystal; (*b*) a small number of grains—notice that even with only three grains the spots begin to appear to form circles; (*c*) a large number of randomly oriented grains—the spots have now merged into rings; (*d*) an amorphous specimen merely gives rise to a diffuse halo, indicating that on average the atoms are similar distances apart.

Two real electron diffraction patterns are shown in fig. 3.16 (*a*). The first was taken from a thin specimen of niobium and shows all the features to be expected of a single-crystal pattern. The lower pattern, also from niobium, has two haloes superimposed on the

(*a*)

(*b*)

Fig. 3.16. Diffraction patterns from two niobium specimens. (*a*) This region consisted solely of a single grain of metal; (*b*) this pattern was taken from a part of the foil which was covered by a layer of oxide. Superimposed on the spot pattern from the metal are haloes from the amorphous oxide film.

single-crystal spot pattern. This tells us that a layer of amorphous material is present as well as the crystalline specimen. In this particular case the amorphous material is a film of niobium oxide which resulted from the method used to prepare the specimen (see Section 3.4 below). The diffraction pattern usefully warned the microscopist that he was not looking at niobium alone but at a sandwich of niobium in niobium oxide.

Now let us look at what can be learned from the diffraction pattern of a single-crystal specimen. We have already seen (fig. 2.12) that the distance, r, between a spot in the diffraction pattern and the centre spot (undiffracted beam) is inversely proportional to the spacing, d, of lattice planes in the specimen. If we can obtain the d spacings of several sets of planes then we can generally deduce the crystal structure of the specimen. Two of the most common crystal structures of metals are shown in fig. 3.17 in crystallographic shorthand. The few atoms shown indicate how the structure repeats itself; the real crystal contains millions of such units. We can see that the two structures shown in fig. 3.17 will give rise to different diffraction patterns. If the crystals were in a different orientation

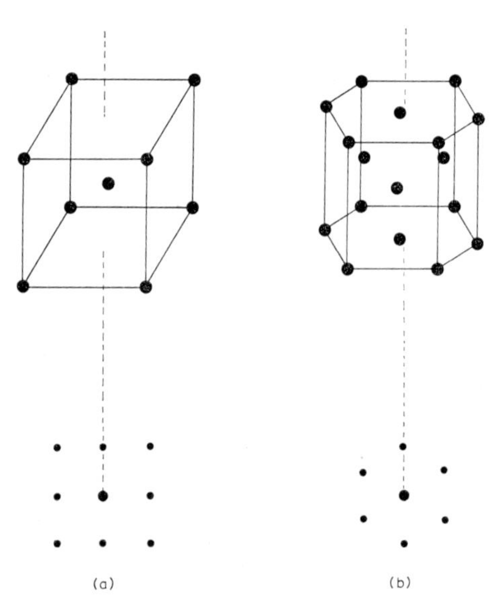

Fig. 3.17. Two of the commonest crystal structures of metals and the diffraction patterns which can arise from them. (*a*) A body-centred cubic (b.c.c.) arrangement of atoms with the electron beam incident on it parallel to the cube edges. (*b*) A hexagonal close-packed (h.c.p.) arrangement of atoms with the electron beam incident parallel to the prism edges.

with respect to the beam of electrons, as shown in fig. 3.18, each one would give a different pattern. It is by combining the information contained in several patterns from the same crystal in different orientations that we can build up a picture of the structure of an unknown crystalline material.

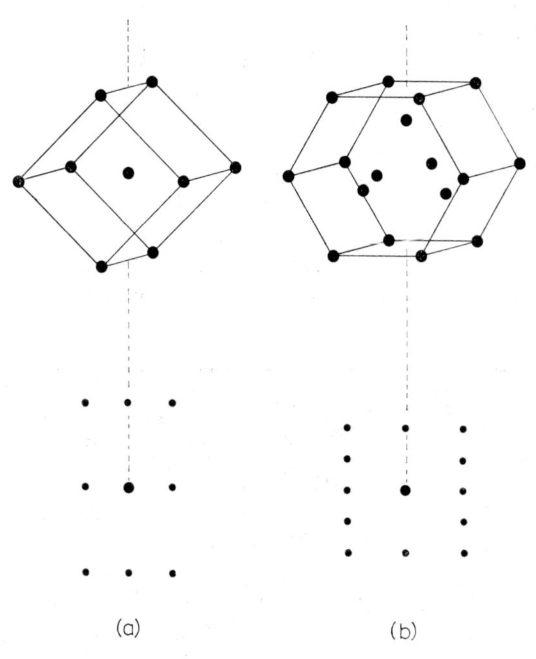

Fig. 3.18. The same two arrangements of atoms as fig. 3.17 with the electron beam incident along a different direction in each. The diffraction patterns are very different from those formed in the orientations of fig. 3.17. Note that pattern (*b*) shows no sign of the hexagonal nature of the crystal structure which gave rise to it.

Often it is simpler to deduce the crystal structure of a specimen from a polycrystalline region because we then have many small crystals in different orientations present at one time and a single ring pattern contains almost as much structural information as several different spot patterns. Thus the ring pattern shown in fig. 3.19 can immediately be recognized as coming from a material with a face-centred cubic crystal structure because of the peculiar grouping of several of the rings in pairs. Frequently, however, we know the crystal structure of our specimen, and we need the spot pattern from a single grain to tell us the exact orientation of the lattice planes in that grain with respect to the electron beam. This is quite easily calculated using vector methods but with a little experience it is possible to recognize familiar spot patterns as they occur in the microscope and hence to know the orientation of each grain as it is observed. The great importance of this lies in the fact that unless we know along which direction in the crystal we are looking we cannot deduce the three-dimensional shapes of any feature which we can only photograph as a two-dimensional projection. The example of helium bubbles in niobium which is described at the end of this chapter should make this point clear. A further reason for needing to know the orientation of each grain which is examined is that the images of defects such as dislocation and small precipitates are very dependent on which set of lattice planes is near the Bragg condition (equation (2.9)). In some orientations of the specimen a dislocation may even be entirely invisible, which could clearly be misleading if it were not appreciated. One complicating factor, which we touched on in Section 2.4, is that since the strength of each lens is altered as we change from viewing the

Fig. 3.19. A diffraction pattern from a polycrystalline specimen of copper. Notice that the rings tend to occur in pairs; this is a characteristic of the face-centred cubic (f.c.c.) structure of this metal.

micrograph to viewing its diffraction pattern the number of turns of the helix which the electrons travel also changes. This effectively rotates the whole image with respect to its diffraction pattern; the photographic plates must therefore be mutually rotated through the appropriate angle before the orientation of the specimen is calculated. The effect of the helical electron paths is easy to see in the microscope—as the magnification is changed (i.e. the strength the projector lens is altered) the image is seen to rotate.

An additional diffraction effect occurs when relatively thick single-crystal regions of a specimen are examined. As well as a pattern of spots in the diffraction pattern the background is seen to be criss-crossed with pairs of parallel lines. The lines, one dark and one light in each pair, are known as Kikuchi lines. The patterns of lines bear a strong resemblance to the electron channelling patterns which can sometimes be seen in the scanning electron microscope (Section 4.4).

Kikuchi lines arise from those electrons which are inelastically scattered in the specimen; it is only in ' thick ' specimens that a large enough number of this type of electron is created. The paths of inelastically scattered electrons (Section 2.5) tend to be very close to that of the incident electron beam (fig. 2.5). Because of this more electrons are travelling along the path AC shown in fig. 3.20 than along AB. Electrons travelling along either of these paths can be diffracted by the lattice planes shown in the diagram and therefore

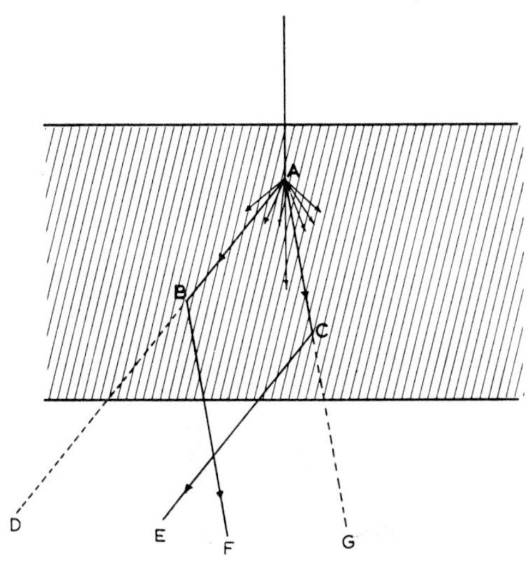

Fig. 3.20. The formation of Kikuchi lines in a thick specimen by Bragg diffraction of electrons which have already been inelastically scattered.

will not contribute to the ' grey ' background of the diffraction pattern in the same way as the majority of inelastically scattered electrons. In the direction AG more electrons will be lost by diffraction to CE than are gained by diffraction into BF. Conversely more electrons are scattered into the direction CE than are lost from BD. Consequently the line in the diffraction pattern to which AG and BF contribute (a line perpendicular to the page) will be darker than its surroundings, whereas the line to which AD and CE are directed will be brighter than its local background. In this way each set of lattice planes can give rise to a pair of Kikuchi lines, one bright and one dark, in the diffraction pattern. These lines occur in addition to the spots normally present, as fig. 3.21 illustrates. The lines are of great interest because their position in the diffraction pattern is very sensitive to the exact orientation of the specimen and therefore the orientation of a fairly thick specimen can be determined very precisely indeed.

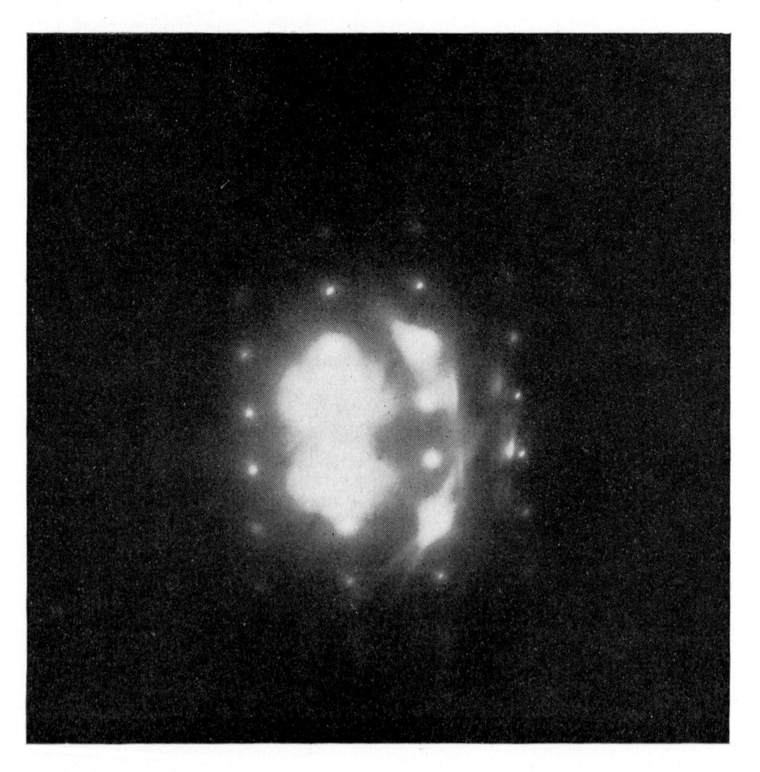

Fig. 3.21. A diffraction pattern from a thin specimen of aluminium showing Kikuchi lines in addition to the spot pattern.

To summarize, electron diffraction patterns can tell the microscopist

(*a*) whether the specimen is crystalline;
(*b*) roughly how many crystals are in the field of view;
(*c*) what the crystal structure is;
(*d*) the distances between planes of atoms in the crystals (i.e. the 'lattice parameter');
(*e*) the orientation of a single crystal (grain) which is being examined.

3.4. *The preparation of specimens for examination in the microscope*

One of the most difficult, yet most essential, aspects of transmission electron microscopy is the preparation of specimens. The requirements are very stringent indeed. In order that the electron beam can penetrate the specimen it must generally be only 100 nm thick or less, which is about one five-hundredth of the thickness of aluminium foil used in the kitchen. Not only must the specimen be made this thin but it must be done in such a way that the structure which we want to see is altered as little as possible. For a crystalline material this means that the specimen must not be bent during thinning or its dislocation structure will be radically changed. For a biological specimen we must ensure that cells or organelles are not torn out of the tissue and that the shapes of distinctive structural features such as mitochondria are changed as little as possible before we view them in the microscope. Since the thinning techniques are very different for biological and non-biological materials we will consider them separately.

3.4.1. *The preparation of biological specimens*

There are two major types of biological specimen, from the point of view of the microscopist; either the specimen is to be looked at histologically, preserving the structure of the tissue so that we can see the relationships between the cells and cell organelles, or the specimen is to be considered from a biochemical viewpoint and one particular component, for example the cell membrane, is to be preserved so that we can look at it in detail even though it may no longer be in its original place in the tissue. Very often the histological approach makes use of 'positive staining' and the biochemical approach needs 'negative staining'. These techniques, and the other necessary procedures (fixation, embedding and sectioning), will now be described briefly.

Fixation is the first essential technique and its aim is to prevent, as far as is possible, any changes taking place in the structure of the specimen after it has been removed from the living organism. Two chemical agents are normally used and both act by cross-linking large biochemical molecules one to another. However, the problem is to introduce the two agents, usually glutaraldehyde and osmium tetroxide, to all parts of the specimen. They diffuse very slowly into a block of tissue and therefore in order to fix a whole specimen in a reasonable time the piece of tissue must be very small. Frequently a cube whose side is only 0·5 mm long is cut. This is not much larger than the full stop at the end of this sentence. Even so it is necessary to soak this cube successively in the two fixatives for several hours. However, after fixation, not only should all the various components be held in their original places but osmium, a heavy metal, is attached to most of the lipids and will act as a stain to enhance the image contrast as discussed in Section 3.2.

The next procedure is to embed the specimen in a resin which supports it while it is handled and then sectioned. The most common embedding medium nowadays is an epoxy resin such as Araldite or Epon. Unfortunately this is not miscible with the water which saturates the specimen after fixation and will not penetrate the specimen tissue until the water is removed. The tissue must therefore be dehydrated by washing it in a series of alcohol–water mixtures with increasing proportions of alcohol until finally the specimen is washed in pure alcohol. This has to be done rather quickly, since if the tissue were left to soak in alcohol some components would dissolve away. Then, almost immediately, the resin is introduced to the specimen. Again, this is done by easy stages; the resin is usually a three-component mixture consisting of resin, hardener and accelerator. The resin and hardener are mixed in the proportions recommended by the manufacturer and a few drops are added to the small vial containing the specimen cube, still soaking in alcohol. After this has been allowed to mix and penetrate the specimen it is poured off and fresh resin and hardener are added. Finally the accelerator is mixed into a fresh portion of resin and hardener and this replaces the earlier mixture. The resin is then left to set around the specimen in a specially shaped plastic capsule; the resin is hardened by being heated in an oven at $\sim 60°C$ for about 48 hours. Finally the plastic capsule is removed, leaving an epoxy block containing the minute cube of tissue as shown in fig. 3.22. As you can deduce from the various times which have been mentioned, it will have taken several days to get the specimen to this stage; the final two processes are rather quicker and may perhaps be done on the same day as the microscopy. From the solid epoxy resin block thin sections must be cut using a delicate instrument called an ultramicrotome. The operation of this instrument is an extremely skilled art; a knife made of diamond or freshly broken glass is made to shave slices from the tapered end of the epoxy block. The slices are generally about 60 nm thick and 1 mm square and their exact thickness

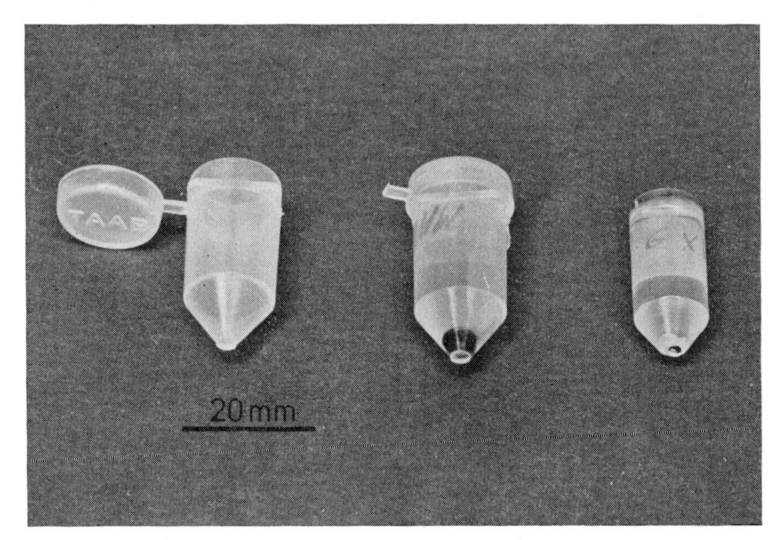

Fig. 3.22. The embedding of biological tissue prior to sectioning with an ultramicrotome. The empty capsule on the left has a small piece of specimen material placed in the bottom of its conical end and is then filled with epoxy resin (centre). After an appropriate curing treatment the solid epoxy block is pushed out of the plastic capsule (right) and the conical end is trimmed to expose the specimen for cutting.

can be judged from their colour as they float on the collecting liquid (fig. 3.23). Sections of about 60 nm thickness appear silvery, whereas much thinner sections are scarcely visible, and those which are too thick to be useful have a purple or bluish tinge.

The section, or series of sections, must now be picked up on a support grid so that it can be mounted in the microscope. This is

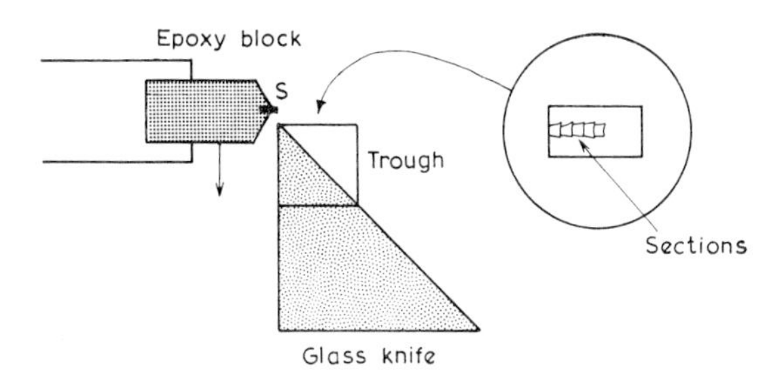

Fig. 3.23. The principle of ultramicrotomy: the specimen (S) in its epoxy block is moved steadily past a freshly broken glass edge (the glass knife). The thin sections float in the liquid-filled trough, each one adjoining the last in a line.

fairly easily done. A 3 mm diameter copper grid can be held in tweezers and brought up underneath the required sections as they float on the surface of the liquid in the trough. As the sections dry they adhere to the grid quite satisfactorily (fig. 3.24).

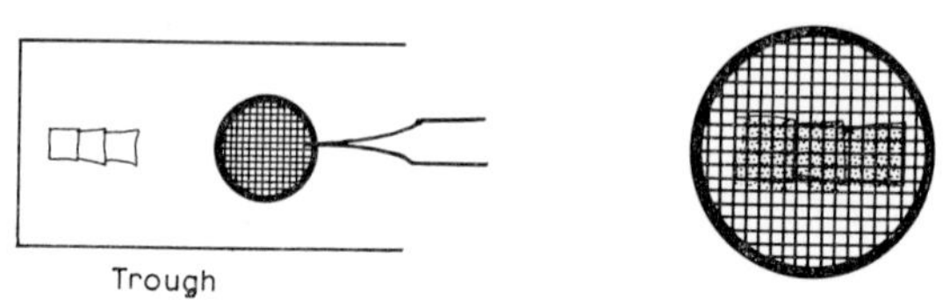

Fig. 3.24. Thin sections are picked up from the surface of the liquid on a copper support grid held in a pair of fine tweezers.

The final action before viewing the specimen in the electron microscope may be to increase the image contrast more than has already been achieved during fixation by staining the section. This is normally carried out by placing the grid and section on top of a drop of staining solution, usually lead citrate or uranyl acetate. After a few minutes the grid and section are removed, washed carefully and allowed to dry on a piece of filter paper. The section is then ready for examination in the electron microscope.

An alternative to the positive staining described above, in which the heavy metal is deposited at certain biochemical sites (mainly fatty radicals), is negative staining, in which the stain aggregates around the small bodies being studied. Negative staining is a technique which has recently been used extensively in the study of particles and fibrous materials such as viruses, bacteria and collagen. The technique is relatively simple: the specimens are mounted on a thin supporting membrane (usually a carbon film) carried on the usual copper grid and a drop of the stain, frequently a solution of phosphotungstate, is placed on the grid and allowed to dry. The difference in action between negative and positive staining is illustrated in fig. 3.25.

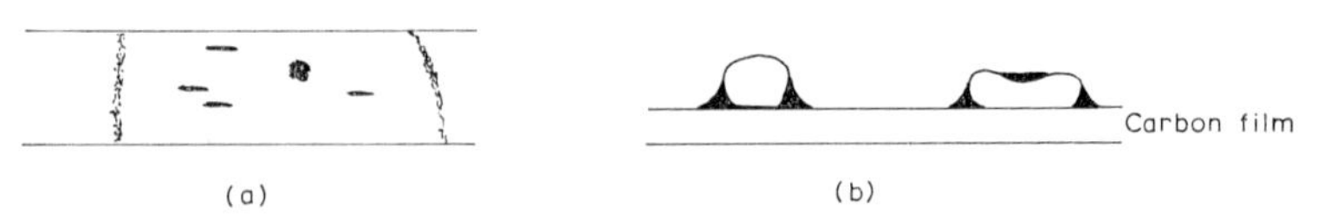

Fig. 3.25. The two types of staining. (*a*) Positive staining, in which the heavy metal penetrates the thin section and is deposited preferentially at particular features; (*b*) negative staining. The specimen particles are supported on a carbon film and then wetted by the staining solution. When this dries the heavy metal collects around the outside of each small feature.

To summarize: the techniques for preparing biological materials for examination in the electron microscope are lengthy and require great skill. Even so, it is possible that the microstructure which is finally seen is not identical to the structure which exists in the living organism. The details which can be observed in a micrograph depend not only on the fixing and staining agents which have been used but may also include features which have been introduced by the slicing action of the ultramicrotome or by the collapse of the section when it is put into the vacuum of the microscope and dries out. It is therefore very important that the microscopist understands the biochemical and physical treatments which the specimen has undergone and that the biologist understands the changes which can take place as the specimen is sliced, evacuated and then bombarded with electrons. It is obviously best if the biologist and the microscopist are one and the same person!

3.4.2. *The preparation of metal and ceramic specimens*

In order to examine specimens of metal, ceramic or other non-biological material we need to prepare a *foil* which is similar in thickness to the biological specimens discussed in the last section, although in some cases useful diffraction contrast can be obtained from specimens more than 100 nm thick. Although such specimens normally have the advantage that they are stable at room temperature and do not need fixing or staining they cannot be prepared by a technique as straightforward as slicing with an ultramicrotome, since they are either too brittle or too susceptible to damage by the bending which inevitably occurs. We therefore have to use entirely different methods and the technique chosen depends very much on whether the specimen is an electrical conductor or not.

The most common technique for electrically conducting materials such as metals and alloys is electropolishing. The principle of this method is that the specimen is made the anode in an electrolytic cell. When a current is passed, metal is dissolved from the anode (the specimen) and deposited on to the cathode. The experimental arrangement can be very simple, as is indicated in fig. 3.26. If the composition of the electrolyte and the operating voltage are chosen successfully a specimen in the form of a thin sheet not only becomes thinner but also smoother and brighter. Eventually a hole appears in the thin sheet and if the neighbouring part of the sheet is sufficiently smooth (i.e. well polished) the regions near the hole will be thin enough for viewing in the TEM. The process is shown schematically in fig. 3.27. The thicker regions of the specimen serve to support the very flimsy thin areas but a small piece of the specimen (perhaps 1 mm square) still needs to be supported in the microscope on a copper grid, as shown in fig. 3.2. There are many variants of the basic

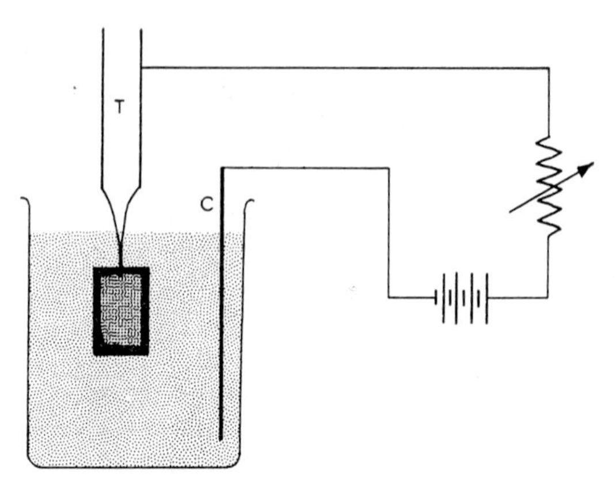

Fig. 3.26. Electropolishing as a technique for preparing thin specimens of metal for transmission electron microscopy. The sheet specimen is held in tweezers (T) in a beaker of electrolyte. A small voltage (1 to 30 volts typically) is applied between the specimen and a metal cathode (C).

electropolishing technique which employ a great variety of different electrolytes. However, the common feature of all the techniques is that they thin a sheet of metal from perhaps 0·1 mm thick down to 0·1 μm (100 nm) in less than an hour. The preparation of a specimen from a thin metal sheet is therefore quite rapid, although if the original piece of metal was thick it may have taken a longer time to machine or grind it to the starting thickness of 0·1 mm.

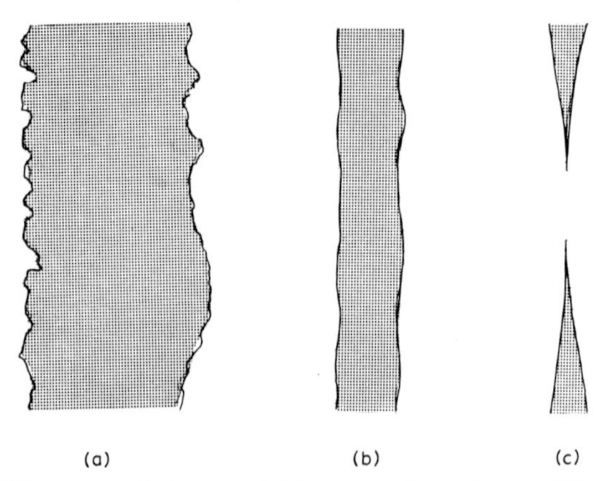

Fig. 3.27. The stages of electropolishing a metal specimen. The thick, rough sheet (*a*) becomes smoother and thinner (*b*) and eventually perforates (*c*). The thinnest regions around the perforation should be suitable for study in the TEM.

Useful though electropolishing is, it suffers from the major limitation that it cannot tackle non-conducting materials. Thus we have to use alternative methods for ceramics such as oxides, for glasses and even for semiconductors such as silicon and germanium. The two major alternatives at present are chemical polishing and ion bombardment. Chemical polishing has proved very successful in thinning many oxides but is not so easy to control as electropolishing since there is no current to switch off when you wish to stop! Most of the chemical polishes simply rely on acid attack but often require rather nasty solutions such as hot concentrated orthophosphoric acid. Until recently, however, this was the only practicable method of thinning many ceramic and semiconducting materials without damaging them.

In the past three or four years a much slower but very reliable method has become widely used. A stream of argon gas is ionized and the ions are accelerated towards the specimen by a potential difference of about 5000 V (5 kV). Each ion, as it hits the specimen, may knock off the surface one or more atoms. This process is known as sputtering and results in the specimen slowly being thinned. With the types of ion source which can be used in an electron microscope laboratory sputtering is extremely slow and may only erode the specimen surface at the rate of about 1 μm per hour. In order to speed up the thinning two ion sources are often used and both sides of a thin specimen are eroded at once in an arrangement such as fig. 3.28. However, even if the specimen has been pre-thinned by some other means to 50 μm (0·05 mm), it will take more than 24 hours

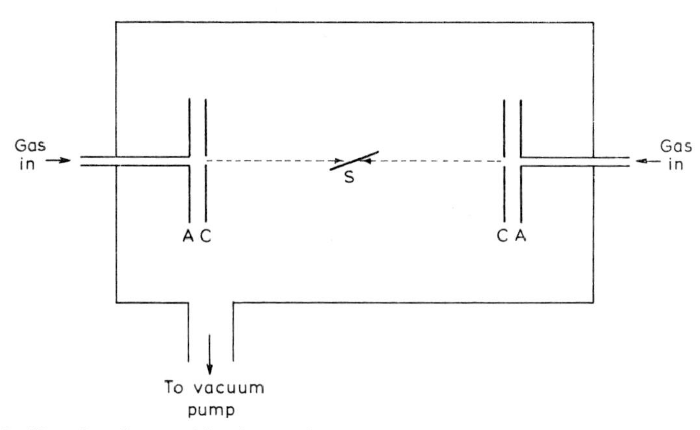

Fig. 3.28. Ion beam thinning. Argon ions are accelerated across a potential difference of several thousand volts applied between the anode (A) and cathode (C) of each ion gun. The two beams of ions are directed at the top and bottom of the specimen (S) simultaneously.

to thin it for electron microscopy. Despite this lengthy time-scale ion bombardment is frequently used, since it is almost infallible (unlike electropolishing or ultramicrotomy) and can handle non-conducting materials.

3.4.3. *Replicas*

There is a further method of looking at both biological and non-biological specimens. This is historically a very early method and consists in making a replica of the surface of a specimen rather than trying to thin the whole piece to electron transparency. This can be done surprisingly easily by depositing a very thin layer of carbon (or a few other materials) by evaporation in a vacuum. The standard technique is to place the specimen in an evacuated bell jar and to heat a carbon rod to white heat by passing a current through it (fig. 3.29 (*a*)). The carbon which evaporates from the solid rod is deposited on all the surfaces inside the bell jar, including the specimen.

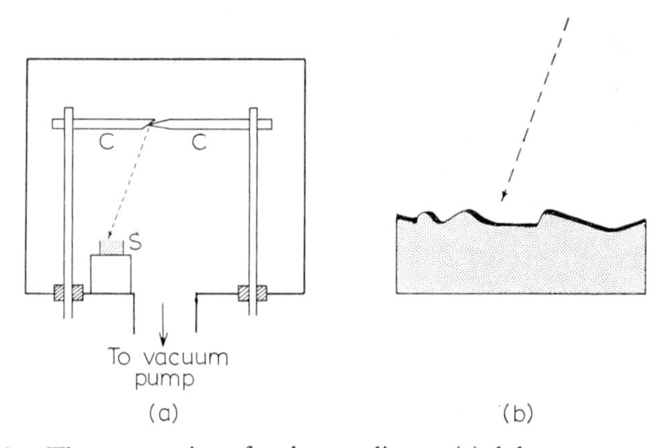

Fig. 3.29. The preparation of carbon replicas. (*a*) A large current is passed through a specially tapered pair of carbon rods (C). Carbon is then evaporated on to the specimen (S); (*b*) if the specimen is offset from the source of carbon, as shown here, the thickness of carbon built up depends on the surface topography of the specimen. The thin carbon replica is then peeled from the specimen surface.

A very thin layer (only a few tens of nanometres thick) is allowed to build up before the evaporation is stopped. The thin carbon film can generally be removed from the surface of the specimen in pieces 1 mm square or larger (although this is often the most difficult part of the procedure) and then mounted on a copper grid for examination in the microscope.

It is perhaps surprising that such a thin film actually retains the shape of the surface of the specimen on which it was laid down. However, this is the case for all but the most complicated surface shapes and the different thickness of material which is deposited on different facets of the specimen (fig. 3.29 (*b*)) gives rise to scattering contrast when the replica is viewed in the electron microscope. An example of a replica micrograph is shown in fig. 3.30. Obviously the technique does not give any direct information about the internal structure of the specimen and therefore replica microscopy is not really an alternative to the thin specimen techniques which we have discussed up to this point. It would seem to overlap much more with the scanning electron microscope technique described in the next chapter. However, the resolution limit (i.e. the smallest detail which we can see) for a replica is about 2 nm, which is rather better than most scanning electron microscopes can achieve (see Chapter 4). There is also a further reason for using replica techniques. Many specimens, particularly alloys, contain two components of different crystal structures. If these two *phases* are very finely divided it becomes impossible to form a diffraction pattern from each phase separately. Consequently the patterns which are seen in the TEM

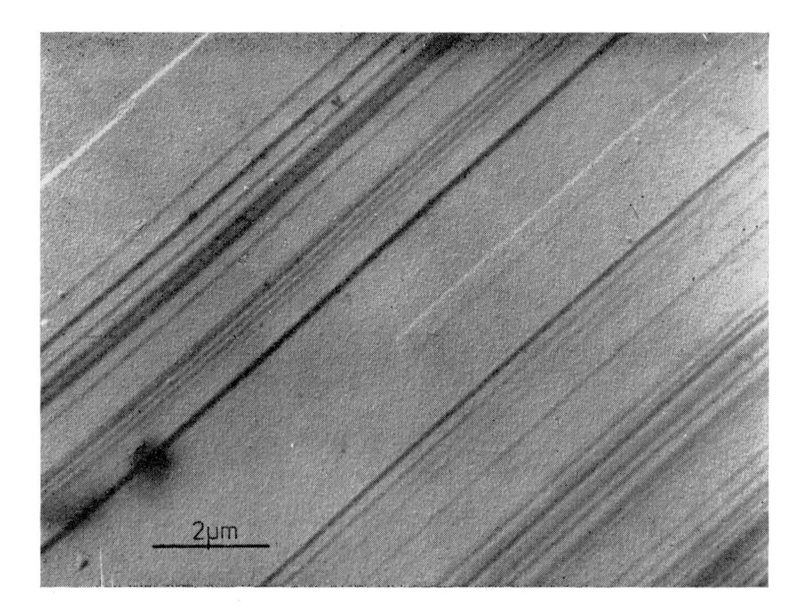

Fig. 3.30. A carbon replica showing slip lines at the surface of a single crystal of aluminium. (P. Charsley.)

are complicated mixtures of the two basic patterns and are very difficult to analyse. Using a replica it is sometimes possible to extract particles of one of the phases when the carbon film is removed. Then a diffraction pattern can be taken from the replica, which will be the pattern from only one phase of the original specimen. This is far easier to interpret. *Extraction replicas* therefore still play a significant part in metallurgical electron microscopy.

3.4.4. *Summary of specimen preparation techniques and possible artefacts*

In order to examine a material in the TEM we must first decide whether it is internal structure or surface morphology which we wish to investigate. This determines whether we are going to thin the specimen itself or make a replica of its surface (or possibly look at it in the scanning electron microscope instead). If we intend to study the internal microstructure we first have to ensure that the features we wish to look at will be visible—for non-metallic materials this may mean staining them. Then we must decide whether to slice the specimen, chemically polish, electropolish or ion-bombard it or whether some other special technique is more appropriate. Finally, when we are looking at the structure in the microscope we must ask " Could that effect have been caused by something which I did to the specimen while preparing it? " Some examples of misleading artefacts are the lines often scored on a thin biological section by slight irregularities in the ultramicrotome knife and the dense dislocation networks introduced near the edge of a thin metal foil as it was accidentally knocked by perhaps one hair of a fine paint brush while being loaded into the microscope!

3.5. *Some applications of transmission electron microscopy*

In order to illustrate the real uses of the instruments and techniques which have been discussed in this chapter I have selected four applications of the TEM to problems in very different areas of science. Each one is described in only sufficient detail to bring out the unique facts which we could learn by electron microscopy but which it would not have been possible to study any other way.

3.5.1. *The effect of virus infection on a chick embryo*

The object of this study was to determine exactly how a virus attacked the individual cells of a chicken embryo. It was decided to examine the rings of cartilage in the trachea of the embryo and to look at embryos at various times after exposure to the infection so that the course of the attack could be followed. However, the first specimen studied, from an apparently non-infected egg, showed signs of virus attack. The micrograph which is shown here (fig. 3.31) is of a specimen from a 20-day-old embryo in which the attack is fairly well developed. This type of specimen can be prepared by absolutely standard techniques: it was in fact fixed with glutaraldehyde overnight and then with osmium tetroxide for one hour. It was then dehydrated in a graded series of alcohols, dried in propylene oxide and embedded in Epon 812. Sections were cut on an ultramicrotome using glass knives; they were then stained with lead citrate and uranyl acetate. The micrograph was taken at an operating voltage of 60 kV.

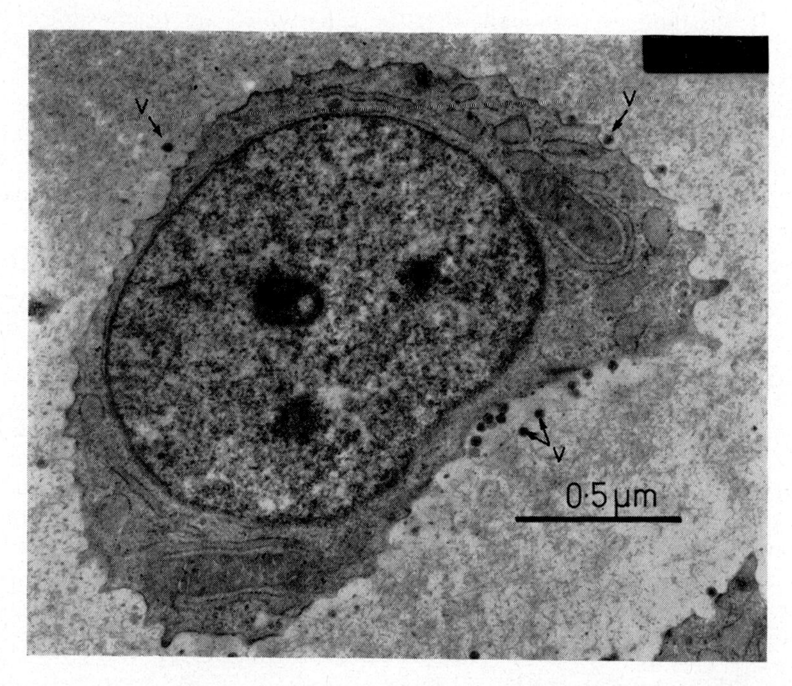

Fig. 3.31. A cartilage cell from the trachea of a chick embryo. The chick has been infected with a virus, which can be seen attacking the cell wall. (N. Abu Zahr.)

Figure 3.31 shows a single cartilage cell. The staining enables us to see the nucleus, the nuclear membrane, the mitochondria and the endoplasmic reticulum very clearly. In addition we can see small circular virus particles scattered just outside the cell which have been stained quite heavily so that they appear very dark. The particular information which this micrograph gives us is that the effect of the virus attack has been to cause the cell membrane to become indented in a large number of places instead of retaining its usual smooth shape. Later micrographs showed that the virus particles entered vacuoles just inside the cell membrane. These, it is thought, then migrate to the nucleus, carrying the virus with them. The TEM study therefore gave us an insight into the way in which virus particles reach the nucleus of a cell. In addition we discovered the unexpected fact that ordinary eggs were already infected with this particular virus.

3.5.2. *The structure of human bone*

It has long been thought that the cells which are responsible for the deposition of calcium in bone become inactive once their initial task is completed. However, recent experiments on astronauts who have been in space for long periods show that they lose calcium from their skeleton. This observation, among others, has led scientists to suggest that the same bone cells which initially extracted the calcium from the blood stream and laid it down as bone could be dissolving bone calcium and returning it to the blood. As part of a programme to investigate this sort of behaviour in more detail the TEM has been used to study cells in adult bone.

Specimens of bone are not nearly as easy to prepare for the TEM as other tissues, so special techniques must be used. In one such technique the bone is partially decalcified using the di-potassium salt of EDTA. This serves two purposes—it is subsequently easier to cut sections from the bone and the collagen fibres contained in the bone are seen more clearly. The standard stages of fixation, drying and embedding are then used as described in the last example. Section cutting with an ultramicrotome is particularly difficult because the bone is not only very hard and brittle but also porous; so a diamond knife is used in a very rigid microtome. The sections are picked up on a specially prepared solution buffered at pH 7·4 so that the calcium distribution is not altered and they are finally stained in the usual uranyl acetate.

The micrograph in fig. 3.32 shows a bone cell in one section prepared in the above way. On the left is a dark region consisting of collagen and calcium, whereas to the right of the micrograph can be seen the decalcified area showing clearly the characteristic bands on the collagen fibres. The striations along each collagen fibre are 64 nm apart and this molecular arrangement is so regular that it would be possible to calibrate the magnification of the microscope using the known spacing. In the centre of the micrograph is the most interesting feature of this specimen. A single osteocyte (bone cell) almost fills the field of view and from it we can learn quite a lot: the most obvious feature is that the cell has two nuclei and the chromatin within them is rather loosely packed. These observations imply that the cell is in a very active metabolic state and not at all ' dead '. Other notable features are that the cell contains rather more mitochrondria than most bone cells but contains no endoplasmic reticulum and is therefore not exporting protein. Also to the right of the cell can be seen a small process, P, containing two unidentified dark objects, which appear to be surrounded by collagen but may in fact be connected to a cell. Obviously these observations must be corroborated by examining a lot more cells in many more bone sections, but this example gives an indication of the vast amount of information which can be deduced from a single micrograph.

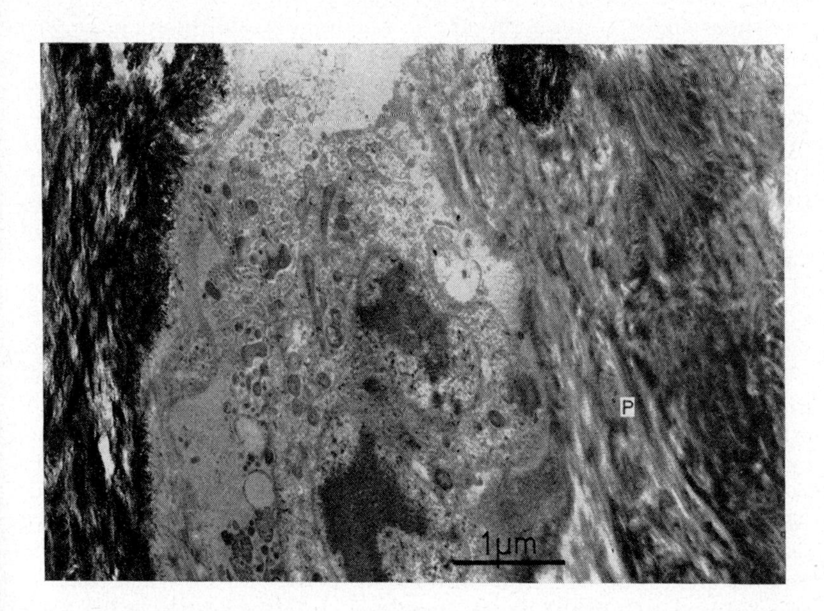

Fig. 3.32. A bone cell (osteocyte) within a region of collagen. (M. J. Dickens.)

3.5.3. *The swelling of metals used in nuclear reactors*

In a nuclear reactor, whether it is in a power station or a submarine, the radioactive fuel must be contained in some sort of vessel or can. Inevitably the materials used in this sort of environment are bombarded with a constant stream of neutrons and other particles emitted by the radioactive decay. The most serious result of this bombardment is that, as well as getting hot, the fuel cans swell and could therefore get stuck inside the reactor, or, worse still, rupture and release their radioactive contents. There are two reasons for the swelling: the neutrons knock atoms out of their normal positions and cause voids to grow inside the metal canning material; and secondly, inert gas atoms are often formed as fission products and deposited in the metal. These aggregate and form bubbles in the metal. It is very important that the mechanism and rate of growth of voids and bubbles are understood so that suitable materials which swell only slowly can be used, and so that the useful life of a fuel can may be estimated.

The bubbles which form in niobium when helium is deposited into it are shown in fig. 3.33 (*a*) and (*b*). After the niobium had been irradiated with helium ions it was heated so that the helium could form bubbles and then electropolished to make specimens for the transmission electron microscope. The bubbles shown are large ones which have been growing for 25 and 100 hours respectively at a temperature of 1050°C. Even so, they are only a few tens of nanometres in diameter and electron microscopy provides the only way of observing them. By measuring the mean size of the bubbles in specimens which have been heated for different lengths of time (0·25 to 100 hours in this experiment) we can determine the rate at which the bubbles grow at this temperature and hence the rate of swelling to be expected. The swelling rate that should be encountered at the temperatures found in the reactor can then be estimated. From experiments taking a few months in the laboratory we can therefore deduce the likely behaviour of the fuel cans in 20 years of service life.

A very noticeable feature of the bubbles shown in fig. 3.33 is that in one micrograph all the bubbles appear to be hexagonal while in the other they all appear square with rounded corners. This behaviour occurs because the bubbles prefer to have certain crystal lattice planes as their faces, so instead of adopting a spherical shape they are faceted. All the bubbles have the same crystal planes as facets and the shape we see depends on the exact crystal direction we are looking along in the microscope. The electron diffraction pattern from each of the specimens shown in the figures enables us to determine exactly which planes form each side of the bubble image and we can therefore build up a picture of the three-dimensional shape of the bubbles. In this case the bubbles form octahedra, which we see as a square, a hexagon or a rectangle, depending on our direction of viewing, as shown in fig. 3.34. This behaviour tells us something not only about the bubbles but also about the physical constants of niobium (in this case its surface energy).

3.5.4. *Phase changes in sintered alumina*

Alumina (aluminium oxide, Al_2O_3) is a strong material which retains a great deal of its strength up to very high temperatures. For this reason, among others, serious consideration has been given to the

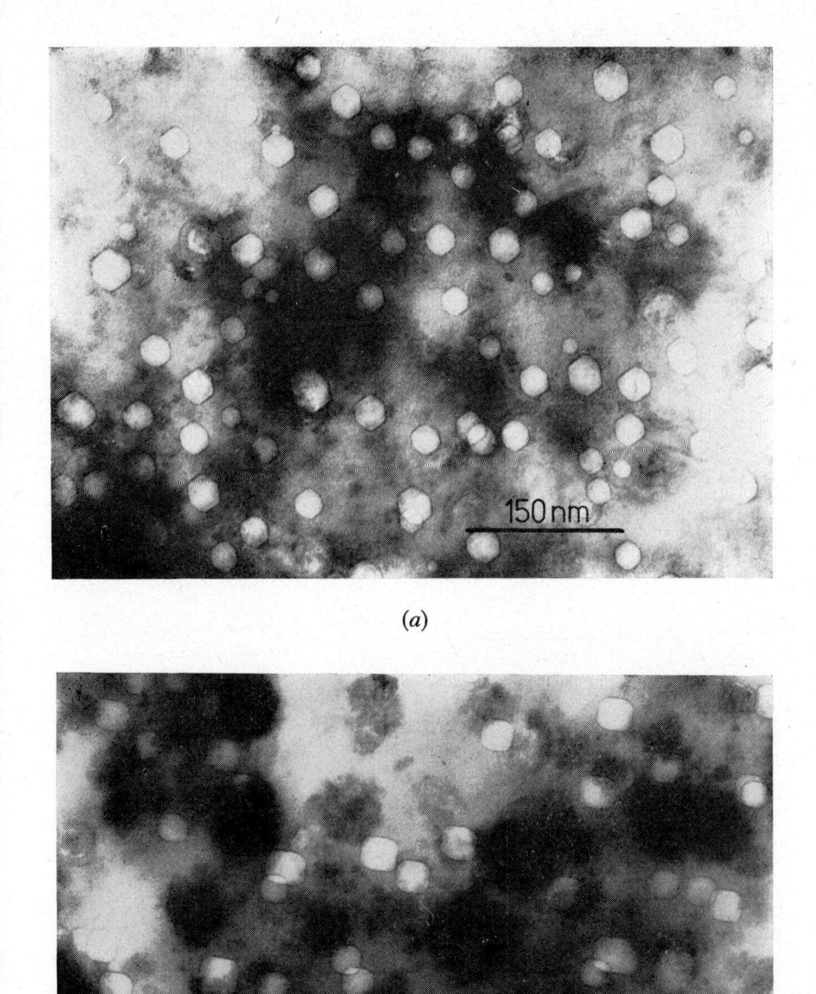

(*a*)

(*b*)

Fig. 3.33. Helium bubbles in niobium which has been irradiated with alpha particles. The shape of the bubble image depends on the direction of viewing, as fig. 3.34 shows, and hence on the crystallographic orientation of the specimen.

possibility of using fine rods of alumina to reinforce gas turbine blades made of nickel alloys. The blades of a gas turbine have to operate at high temperatures and they must be very stable, since there is very little clearance between the blades and the casing of the turbine, and as light as possible. The addition of rods of alumina to the nickel alloy should be beneficial since it should reduce the weight of the blades while giving them added strength and rigidity.

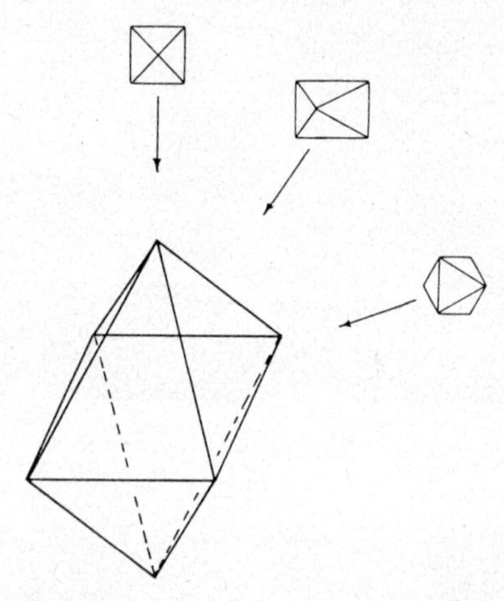

Fig. 3.34. The bubbles shown in the previous figure are in fact octahedral. They will appear to be square if viewed directly down on to the apex, hexagonal if viewed directly on to the centre of a face, and rectangular in any other orientation.

However, it is not easy to produce strong rods of very fine diameter from the powdered form of alumina which is generally available. It is necessary to 'sinter' the powder by heating it to a temperature well in excess of 1000°C while it is held in the form of a fine rod. To achieve this on an industrial scale the rod must be heated and subsequently cooled rather quickly and the 'grain size' of the alumina must remain very small. Unfortunately alumina can exist in at least six different crystalline forms and it happens that the heat treatment necessary for sintering encourages transformation of the desirable form (θ-alumina) into much larger grains of a different type (α-alumina). In order to discover exactly how this happens and therefore how it could be discouraged a transmission electron microscope study of the system was undertaken.

The preparation of alumina specimens for TEM study is rather difficult. Alumina is a non-conductor and therefore cannot be electropolished; also these particular specimens were very porous and broke up if the usual method of mechanical polishing was attempted. The only practicable method was ion-beam thinning and this is how the specimen shown in fig. 3.35 was prepared. In this micrograph it is possible to see the fine-grained polycrystalline θ-alumina region and the coarser region which has transformed into the α form. Although many large pores are still present the electron diffraction patterns showed that the α region is in fact a single crystal and therefore much weaker than the polycrystalline phase. From a series of specimens heat-treated for different times the rate at which the transformation occurred could be deduced. The unfortunate end to a successful piece of microscopy was that no way could be found to make the alumina rods strong enough and cheap enough for the desired application!

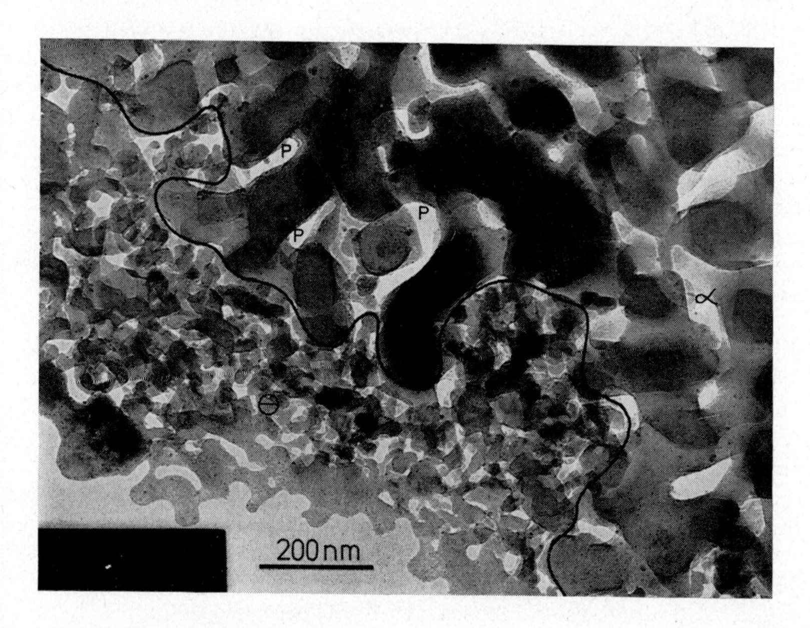

Fig. 3.35. A very porous specimen of sintered alumina. The pores (P) are very evident, as is the division of the structure into the fine-grained θ-phase and the coarser α. (P. A. Badkar.)

CHAPTER 4

the scanning electron microscope

4.1. *How it works*

THE scanning electron microscope (SEM) is similar to the transmission electron microscope (TEM) in that they both employ a beam of electrons directed at the specimen. This means that certain features are the same—for instance the electron gun, the condenser lenses and the need for a vacuum system—but there the similarity ends. The SEM looks at a very different aspect of the specimen and it uses an entirely different method of magnifying the image. However, let us first emphasize the similarities to the TEM. A diagram of the electron optical system of a typical SEM is shown in fig. 4.1, and fig. 4.2 is a photograph of a modern instrument. An electron gun, usually of the tungsten filament thermionic emission type, produces electrons and accelerates them to an energy between about 2 keV and 50 keV, rather lower than typical TEM energies (20–100 keV). Two or three condenser lenses then demagnify the electron beam until, as it hits the specimen, it may have a diameter of only 5–10 nm.

The fine beam of electrons is scanned across the specimen by the deflector coils, D, while an electron detector counts the number of low energy secondary electrons (see fig. 2.7) given off from each point on the surface. At the same time the spot of a cathode-ray tube (c.r.t.) is scanned across its screen while the brightness of the spot is controlled by, for instance, the number of electrons counted by a secondary electron detector (fig. 4.1). The beam of electrons and the c.r.t. spot are usually both scanned in a similar way to a television receiver, that is in a rectangular set of straight lines known as a *raster*. The mechanism by which an image is magnified is then extremely simple and involves no lenses at all. The raster scanned by the electron beam on the specimen is made smaller than the raster displayed on the c.r.t. The linear magnification is then the side length of the c.r.t. divided by the side length of the raster on the specimen (fig. 4.3). For example, if the electron beam is made to scan a raster $10\ \mu m \times 10\ \mu m$ on the specimen and the image is displayed on a c.r.t. screen $100\ mm \times 100\ mm$ the linear magnification will be $100\ mm/10\ \mu m = 10\ 000\times$. With present-day electronics it is simple to construct deflection coils and their controls so that the electron beam can be made to scan a raster as small as $1\ \mu m \times 1\ \mu m$; magnifications of $100\ 000\times$ and above are therefore easily attainable.

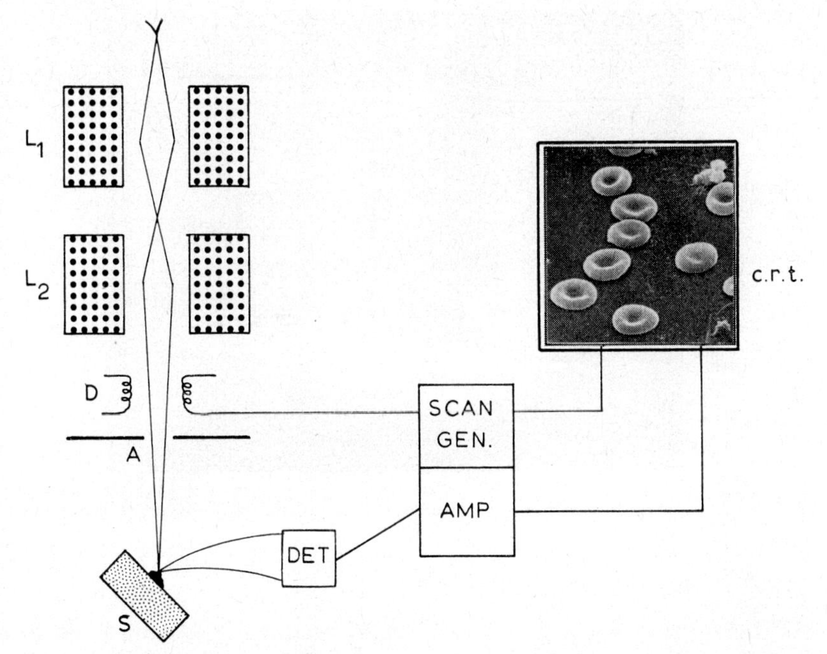

Fig. 4.1. A schematic diagram showing the operation of a scanning electron microscope. The electron beam is focused on to the scanned specimen (S) by the condenser lenses (L_1 and L_2) and across it by the deflector coils (D). The secondary electrons are converted into a current by the detector; this current is then amplified and used to control the brightness of the cathode-ray tube (c.r.t.). The image shown here is of red blood cells.

However, as we saw in Chapter 1, magnification is not a very useful criterion of a microscope's performance; we should be discussing the resolution of which the system is capable. The ultimate resolution, which we defined in Chapter 1 as the smallest separation of two points which the microscope can detect as being separate entities, is determined largely by the diameter of the beam of electrons which is scanning the specimen. Figure 4.4 illustrates this point: on the left is a sketch showing the path of the electron beam across the specimen. For the best resolution the spot has been made as small as possible (5–10 nm in diameter for a typical SEM) and each line of the raster is adjacent to the previous line. Only three lines are shown, as they pass over a region of the specimen containing five small features. On the right of fig. 4.4 are shown the equivalent three lines of the cathode-ray tube display. Here the scale is much grosser; the spot is about 0·1 mm in diameter, so we can see it easily on the screen.

Fig. 4.2. A typical scanning electron microscope, the Stereoscan (Cambridge Instrument Co.).

As the electron beam passes each of the features A, C and E on the specimen the detector receives an increased number of secondary electrons and therefore the spot on the c.r.t. becomes brighter at these points (we have shown it darker in the diagram for clarity). On its next pass the electron beam only touches D and E and thus two bright dots appear on the c.r.t. On the third pass only B is detected. In terms of resolution we can interpret this in the following way.

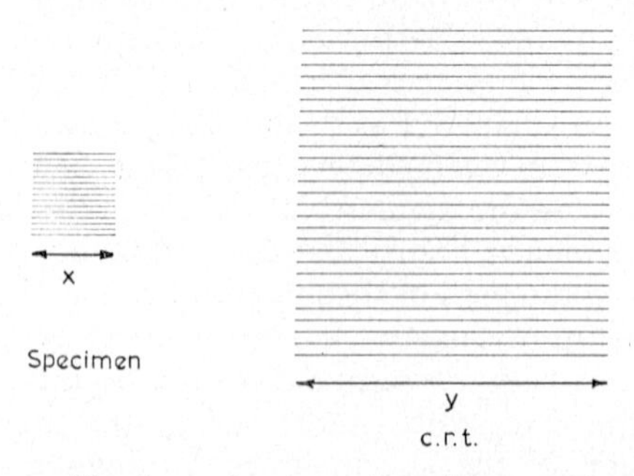

Fig. 4.3. The magnification of a scanning electron microscope; if a small raster of side length x is scanned on the specimen by the electron beam. while a larger raster of side length y is scanned on the c.r.t. the magnification will be y/x.

Features A and B were farther apart than the diameter of the electron beam and gave rise to two distinct dots on the display c.r.t. Features C and D, however, which were only separated by a distance equal to or less than d_0 give rise to a slightly elongated dot which cannot be distinguished from the dot representing the feature E. Points C and D are therefore not resolved and we can say that the geometrical resolution limit is d_0, the diameter of the electron beam. In practice the electron beam excites secondary electrons from a region a little larger than d_0 and therefore the attainable resolution is a little worse than d_0. In modern SEMs a beam diameter of 5–7 nm is used to give a resolution of better than 10 nm.

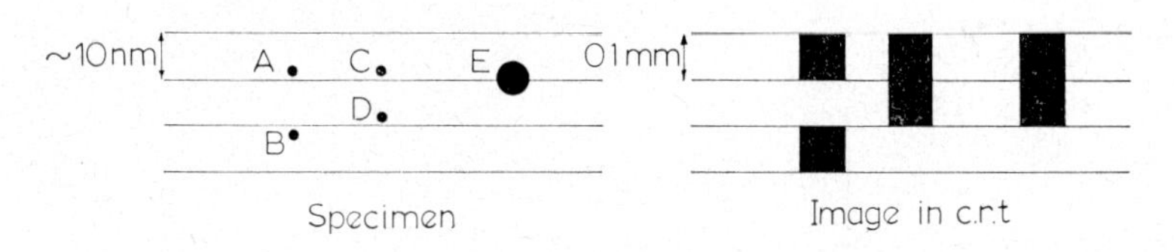

Fig. 4.4. The concept of resolution in the scanning electron microscope. The small features A and B are resolved, since their separation is greater than the electron beam diameter, whereas the two features C and D cannot be distinguished in the image from the single larger feature E.

The factors which limit the diameter, d_0, are many, but in most SEMs two considerations are dominant and both involve the size of the final aperture (A in fig. 4.1). The most important point is that enough electrons must be incident on the specimen per second for the secondary signal to be big enough and representative enough to form an image in a reasonable time. The implication of this is that as we make the beam diameter smaller we must make the final aperture larger. The aperture is usually defined in terms of its semi-angle, α, in the same way as it was for the light microscope and the TEM (fig. 1.9). We can then plot the necessary aperture against the desired beam diameter as shown by the line labelled d_0 in fig. 4.5. Unfortunately, we cannot use as small a beam diameter as we would like just by increasing the size of the final aperture. As the aperture size is increased the spherical aberration of the final lens becomes important and the beam diameter is limited to the circle of least confusion (Chapter 1) which increases very rapidly with α. The effect of this aberration is shown in the line labelled d_a in fig. 4.5. The combination of the two effects is shown as the full line in fig. 4.5 and it indicates that there is an optimum value of α which is needed to get the smallest beam diameter. Unless a fundamental change can be produced in the number of electrons leaving the electron gun there is no possibility of reducing the smallest beam diameter, and hence the resolution of the SEM, below d_{min}.

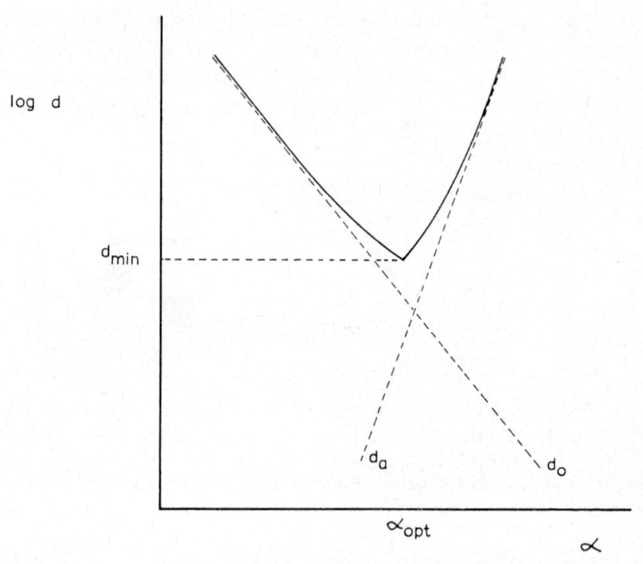

Fig. 4.5. The variation of electron beam diameter with angular aperture α for a particular value of beam current. As the aperture is made larger, admitting more electrons, the beam can be used at smaller diameters (dotted line d_0). At larger aperture sizes however spherical aberration increases the beam diameter (dotted line d_a) and there is therefore an optimum aperture α_{opt} which permits the smallest beam diameter d_{min} to be used.

Variation of the size of the final aperture affects not only the resolution of the SEM but also its depth of field (defined in Section 1.5). Unfortunately the large apertures necessary for the highest resolution do not lead to a very great depth of field: in general the smaller the final aperture the better is the depth of field. Figure 4.6 should make this clear: Δ is the distance along the axis of the microscope over which the diameter of the beam does not change detectably. The actual value of Δ will depend on the beam diameter and the magnification at which the image is viewed but it is evident that the smaller we can make 2α the greater will be the depth of field. To decrease 2α we can either use a smaller aperture or move the specimen farther away (i.e. increase the working distance, W.D.).

Both of these courses of action will make the resolution of the SEM worse, so in practice the size of the aperture has to be chosen either to enhance the depth of field *or* the resolution, but not both. However, even in conditions of ' poor depth of field ' the value of Δ is much better than can be obtained with a light microscope because α is still very small.

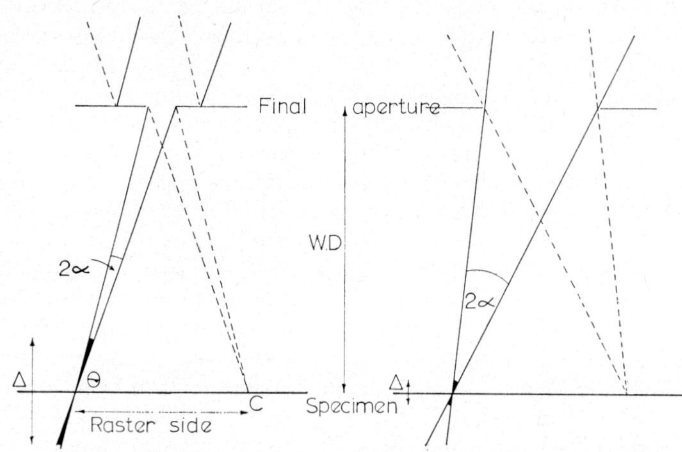

Fig. 4.6. The effect of aperture size on the depth of field at a given working distance (W.D.). The smaller aperture, on the left, gives rise to the larger depth of field Δ.

The scanning electron microscope clearly differs from the transmission microscope in one extremely important way: in the SEM the formation of a magnified image is achieved without the use of lenses. This gives it two significant advantages—the image cannot suffer the aberrations which are unavoidable when using lenses and also the image information (the signal) is carried as a current in a wire and it can, therefore, be amplified and electronically manipulated in a wide variety of ways, as we shall see in a later section. On the other hand, compared with a TEM the resolution of the SEM is rather poor, being limited by the minimum diameter of the electron beam. We shall see in Section 4.3 that use of the field emission electron gun is one of the few ways of improving the resolution appreciably.

The discussion so far has concentrated on the maximum possible resolution, which implies that the SEM is used at high magnifications (above 10 000 ×). However, the great advantage of the SEM is not solely its resolving power but the enormous depth of field which is available when viewing a specimen. This arises from the small angular apertures usually used (α in equation (1.5)) and is usually about 1000 times better than for a light microscope working at a similar magnification. You will find, for example, that many of the micrographs illustrating applications of the SEM in the last section of this chapter have been taken at fairly low magnifications. Such magnifications are easily within the capability of a good light microscope but in that case such a poor depth of field would be available that most of the specimen would be out of focus.

Let us therefore consider the operation of an SEM at low magnifications. The spot on the c.r.t. must stay approximately 0·1 mm in diameter and this implies that to fill a screen 100 mm × 100 mm we must have 1000 horizontal lines. Consequently we must scan the specimen area being examined with 1000 lines. However, if we retain the electron beam diameter as small as we needed it for highest resolution (i.e. $\sim$10 nm) we shall be scanning only a minute portion of the whole surface. For example, to examine a specimen at 100× we must scan an area 1 mm × 1 mm, which means that each of the 1000 lines has a 1 μm strip available to it. If the beam diameter is only 10 nm then only one-hundredth of each strip (10 nm/1 μm), and therefore one-hundredth of the whole area, is really examined. For this reason it makes more sense and gives a bigger signal (which is easier to display) if the electron beam diameter is increased to nearer 1 μm. This is easily done by altering the power of the condenser lenses, and the normal practice is to choose a beam diameter appropriate to the magnification of the image.

To summarize the working of an SEM, the surface of the specimen is scanned in a raster by a fine electron beam, and the variation of any secondary effect—electrons usually—is studied on a cathode-ray tube or TV screen. For reasons explained in the next section this gives an image of the surface of the specimen. It is rather difficult to present the image on a TV screen with no flicker, especially at high magnifications. Consequently it is often necessary to view a c.r.t. on which the spot takes a second or two to travel from top to bottom. This results in the appearance of a line travelling down the screen every few seconds but otherwise causes no difficulties.

4.2. *Why we see anything*

The discussion in the previous section has shown how any variation in the number of secondary electrons reaching the detector as the electron beam scans the specimen can be presented on a much larger scale as the variation in brightness of a spot on a cathode-ray tube. Why should this ‘ variation in brightness ’ form an image of the specimen surface which very often we can recognize immediately? To answer this question we must consider carefully what our eyes recognize when they view the surface of an object. The simple answer is that they see whatever light is reflected towards them from the surface of the object. The first necessity, therefore, is that the object must be illuminated with light and this might be direct sunlight, giving an intense illumination from one direction, or it might be diffuse daylight reaching the specimen from many directions after being reflected from clouds, buildings, etc. Our eye then distinguishes detail in the object either because parts of it reflect a great deal of the illumination while other parts do not or because one colour of light is reflected preferentially from one area and another colour from elsewhere. An example of the first type is the print in this book—the page reflects a great deal of light (of all wavelengths) while the ink scatters or absorbs it and, therefore, appears black. The other type of vision, involving colour, is of less interest to us, since for the present the SEM can only ' see ' in black and white. However, there is a vast amount of detail visible in black and white and to illustrate how and why both the SEM and the eye detect this detail I suggest you take a small coin and place it, head upwards, on the table in front of you.

If you look at the coin from a distance of 30 to 40 cm and the light in the room is not shining directly at it you will probably be able to make out that a head and some lettering is embossed on the coin. You may recognize the head, since you know what to expect, but I doubt if you will be able to read the lettering. If you now pick up the coin and hold it so that it catches the light, so that the room light (or sunlight) is reflected at your eyes by the coin, you will be able to see the detail much more clearly. Notice which areas appear bright and which appear dark; the flat areas such as the background, the cheeks, neck and shoulders of the head and the main part of each letter generally appear bright while the complicated patterns in the hair and, particularly, the outlines of the head and each letter appear dark. The reason is that the flat areas are oriented to reflect the light straight at your eyes, whereas the outline of the embossed areas consists of metal which is very differently oriented and which, therefore, reflects the incident light elsewhere so that very little reaches the eye. In the diffuse illumination conditions when the coin was lying on the table light was incident on the embossing from all angles and therefore no particular feature, whatever its angle to your eye, was reflecting very much more light to you than any other part of the coin. Therefore you saw very little *contrast* between one part of the coin and another. We shall now see that exactly the same considerations determine what we see in the SEM.

Figure 4.7 illustrates the very strong analogy between the information about a solid specimen detected by the eye and that detected by the SEM. Figure 4.7 (*a*) shows a surface under diffuse lighting conditions. Light is arriving from almost all directions so that whatever the orientation of each of the facets A, B and C some light is reflected towards the eye. Figure 4.7 (*b*) shows the same surface illuminated with a parallel beam of light; only the A facets are now correctly oriented to reflect light to the eye. Facet C reflects the light away from the eye, while facet B cannot even be reached by the illuminating beam. The analogous situations in the SEM are shown in fig. 4.7 (*c*) and (*d*). The only major difference is that the electrons are travelling in the opposite direction to the light rays. Figure 4.7 (*c*) illustrates the situation in the SEM when the electron detector has a positive voltage applied to it and therefore attracts low energy secondary electrons. Consequently two types of electron are ' seen ' by the detector; those with very high energy, the reflected primary electrons shown travelling in straight lines in the figure, and the very low energy secondary electrons which follow curved paths as they are dragged towards the detector. There are usually far more secondaries than reflected primaries so we normally refer to this imaging mode as the secondary electron mode. Because the low energy (and hence low velocity) electrons can be deflected round corners the detector ' sees ' electrons from all the facets A, B and C. The electrons from B have the best chance of avoiding the detector but even so this region of the specimen should not look much less bright than the rest, just as in the case of straightforward viewing with the eye. However, if the electron detector has a negative voltage (perhaps 200 V) applied to it all the low energy secondary electrons, which have energies of less than 50 eV, will be repelled and the only electrons to be ' seen ' by the detector will be the high energy reflected primaries as indicated in fig. 4.7 (*d*). The two consequences of using this reflected mode of viewing are that there are far fewer electrons reaching the detector, so the ' signal ' is weaker, and the detector can no longer pick up electrons (and hence information) from round corners. Consequently, just as in the case of parallel light illumination, we can no longer see facet B. The SEM therefore behaves exactly as our eye as long as we remember that the image will appear as if our eye was looking along the electron beam and the specimen was illuminated by light from the electron detector.

The appearance of a ½p coin in these two SEM imaging modes is shown in fig. 4.8. It shows up an important difference between the eye and the SEM. Although the micrograph taken using secondary electrons shows the embossing in much less contrast than the micrograph taken with reflected primaries it also shows a dappled dirty-looking surface. The reflected primary micrograph on the other hand looks very clean. To understand this we have to consider the origin of the electrons which reach the detector in each case. The secondary electrons, which are in a majority in the first micrograph, have such low energies that they can only escape from within 10 or 20 nm of the specimen surface. They are therefore

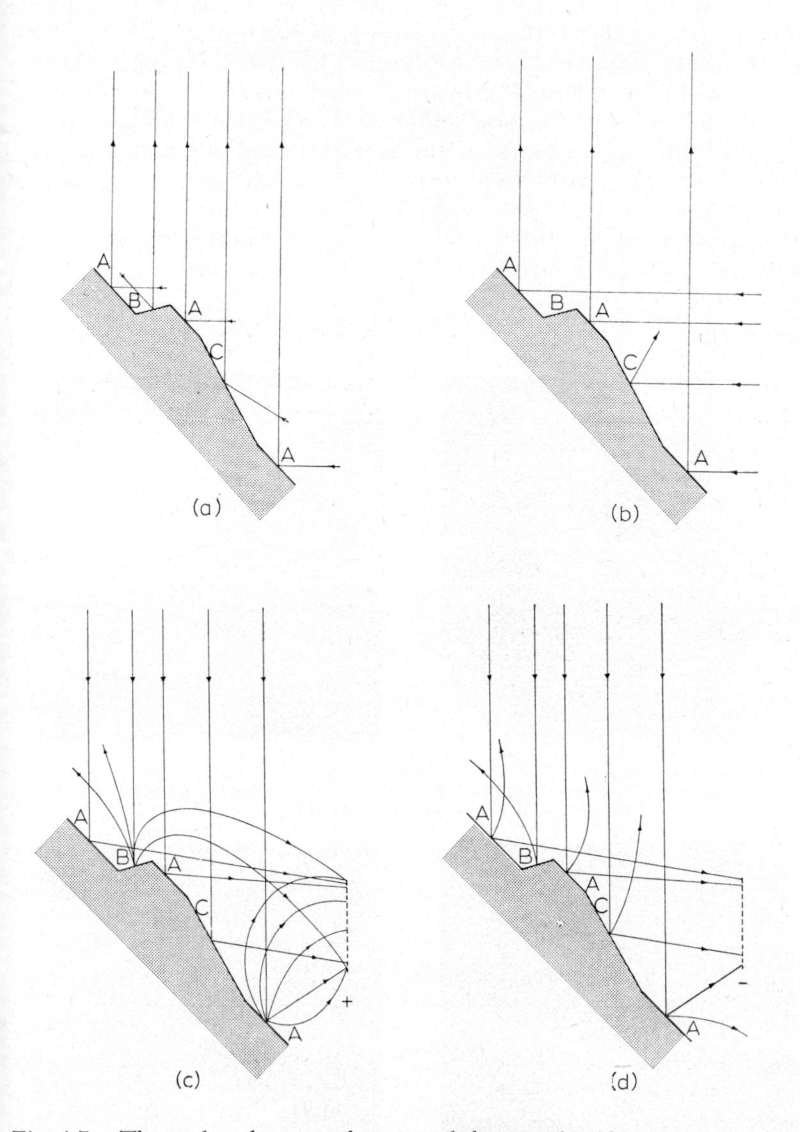

Fig. 4.7. The analogy between the eye and the scanning electron microscope. The four diagrams show the same specimen viewed under different conditions. (*a*) Diffuse illumination viewed by the eye (looking down from the top of the page); (*b*) direct illumination viewed by the eye; (*c*) the equivalent of (*a*) in the SEM, secondary electrons being detected; (*d*) the equivalent of (*b*) in the SEM with only reflected primary electrons being detected.

heavily influenced by the exact nature of the surface of the coin, which will be stained with grease from many a sweaty palm. On the other hand, the reflected primaries, which are in the majority in the second micrograph, have sufficient energy (tens of kilovolts) to emerge from much deeper in the specimen and although they are sensitive to the orientation of the specimen they are relatively unaffected by the exact nature of its very surface. In the case of the coin shown in fig. 4.8 there is little doubt which image has the greater aesthetic appeal, although we should beware of choosing illustrations solely on these grounds!

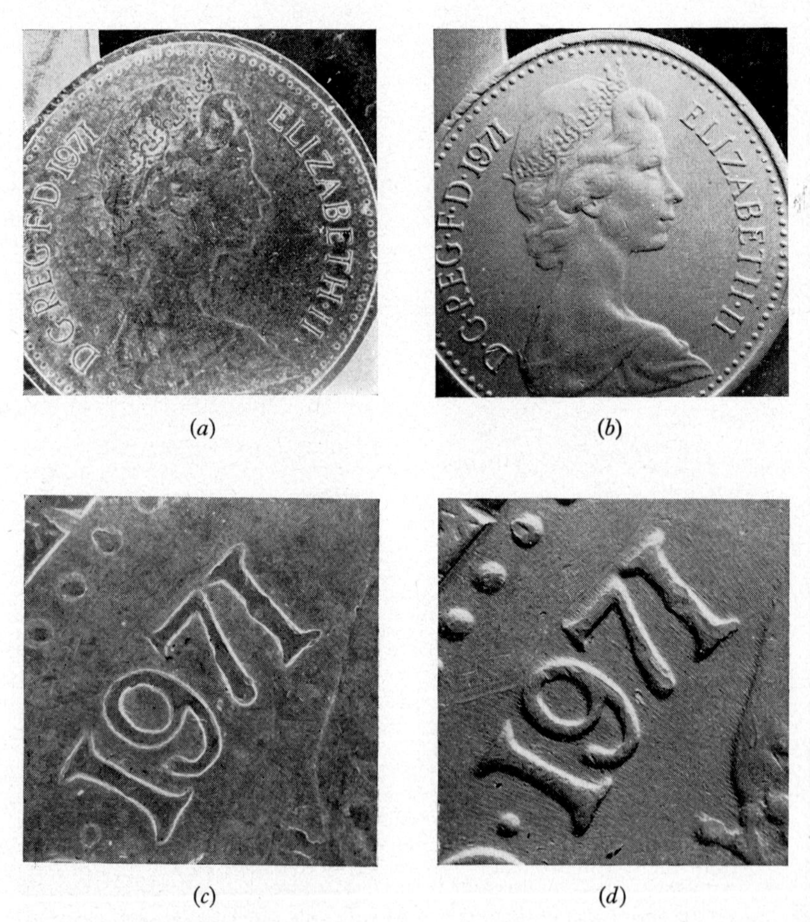

Fig. 4.8. Details of a $\frac{1}{2}$p coin taken using a scanning electron microscope in the secondary mode (*a*) and (*c*) and in the reflected primary mode (*b*) and (*d*).

Before leaving the topic of contrast we should make clear a difference between the eye's method of viewing and the SEM's, which is glossed over in fig. 4.7. In fig. 4.7 (*a*) and (*b*) the eye sees all the specimen surface at one time. This is, of course, not true in the case of the SEM and in fig. 4.7 (*c*) and (*d*) we must remember that the information from each point is collected consecutively. The five beams of electrons shown in the figure would not be hitting the specimen at once; rather they are five positions of the same electron beam at different moments. This does not, however, invalidate any of the considerations of image contrast which we have just discussed.

4.3. *The variety of imaging modes possible with an SEM*

We have so far assumed that a scanning electron microscope forms an image by using the electrons given off from the specimen surface. However, one of the great strengths of the SEM system is that *any* secondary effect which is excited by the incident electron beam can be used to provide the signal which controls the c.r.t. brightness and hence the image. The most commonly occurring secondary effects have already been summarized in fig. 2.7; any of these effects can be detected and converted into an electric current. Consequently in addition to the secondary electron mode and the reflected electron mode which were described in the last section we can also form images in the X-ray mode, in the transmission mode or in the luminescent mode if we use the appropriate detector. In addition there is a less obvious effect which can be used—the specimen current. In most circumstances the number of electrons leaving the specimen at any moment is not the same as the number incident on the surface. This must result in a charge building up in the specimen unless a current is allowed to flow to or from earth. We can amplify this current and use it to display a conductive mode image of the specimen.

We shall meet the X-ray mode of imaging when we deal with electron probe microanalysis in the next chapter. Let us consider here what information about the specimen we can gain from the other three modes. The conductive mode would appear at first sight to give us a negative picture of the surface. Where more electrons are given off, leading to a brighter secondary image, fewer are conducted to earth, giving a darker conductive image. For many specimens this is true and therefore the conductive mode offers no advantage over the secondary mode. However, for the examination of semiconducting specimens such as integrated circuits the technique comes into its own. To understand why, we need to look more closely at what happens inside our specimen when the electron beam hits it. In Chapter 2 we summarized those effects generated by the incident beam which we could detect *outside* the specimen. These represent only the tip of the iceberg. Every incident electron generates hundreds or even thousands of ' electron–hole pairs ' when it knocks electrons out of the outer shells of the atoms of the specimen, giving rise to a free electron and a ' hole ' in the outer shell. Normally the vast majority of these pairs recombine within about 10^{-12} seconds—in other words the electrons jump back into their places in the shells extremely quickly. However, if the specimen is a semiconductor and a potential difference (voltage) is applied across it the electrons and holes will be dragged apart before they can recombine and a current will flow between the electrodes (fig. 4.9). If we use this specimen current as our SEM signal we can display an image representing the variation of some very interesting semiconductor properties across the specimen. Briefly the specimen current flowing from each point will be influenced by the conductivity of the specimen at that point, the lifetime of the electrons and holes (the average time before they recombine), and their mobility (the drift speed under unit potential gradient). These three parameters are of extreme importance to the semiconductor industry and if their variation across a specimen can be made visible in the SEM then faults such as impurities, poor contacts, etc., can be spotted within a single integrated circuit without damaging it. It is even possible to study an integrated circuit chip, almost too small to be seen with the naked eye, while it is being used, with currents flowing through its various components. Figure 4.10 shows an example of this.

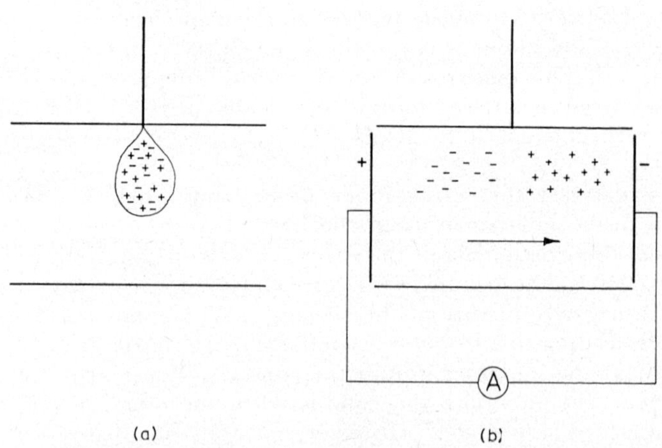

Fig. 4.9. The ' charge collection current ' mode of imaging in the SEM. (*a*) Electron–hole pairs are generated within any specimen; (*b*) if a potential difference is applied across the specimen the electrons and holes can be separated before they recombine and a current will flow in the external circuit. This current is used as the signal for the formation of the SEM image.

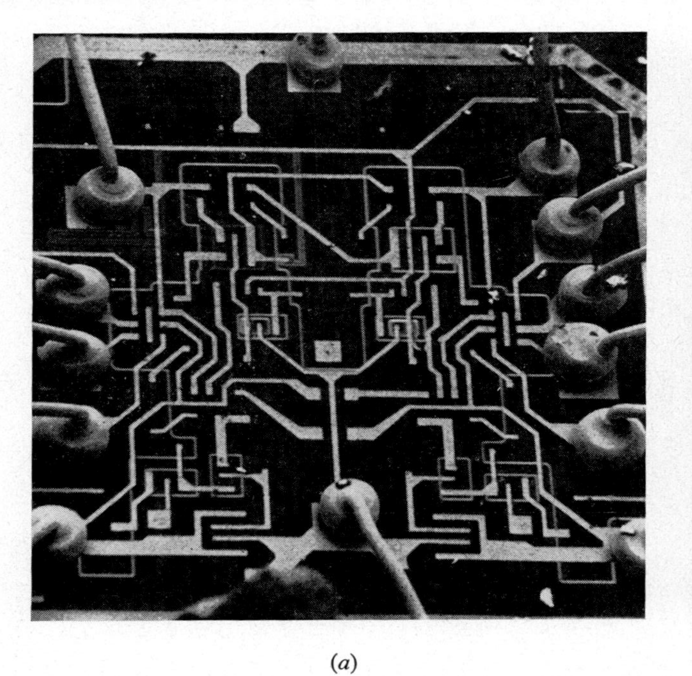

(*a*)

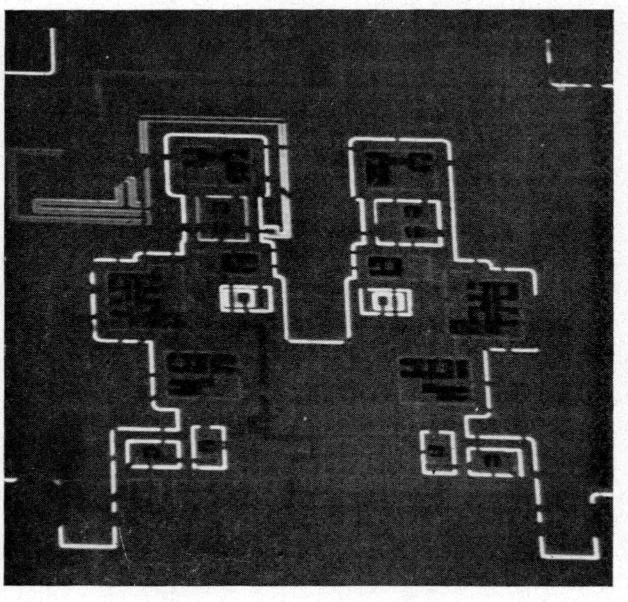
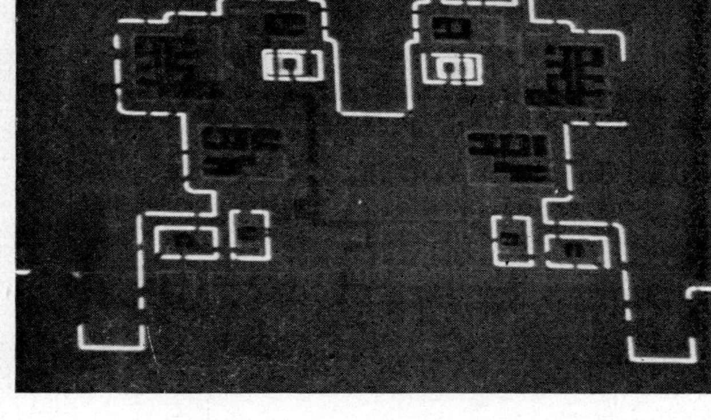

(*b*)

Fig. 4.10. Two micrographs of the same integrated circuit. (*a*) Using secondary electrons to show up the surface features such as the contact wires; (*b*) using the specimen current to show the electrical state of some of the components below the specimen surface. By altering the electrical supplies to the circuit different components can be made to ‘ light up ’. (G. J. Hill.)

Many materials emit light under electron bombardment, and if this is detected we can display an image in the *cathodoluminescent mode*. We have already encountered one such material, the phosphor used on the viewing screen of a transmission electron microscope, and we constantly see others on the c.r.t. of the SEM or on the tube of a domestic television set. Cathodoluminescence varies in colour and intensity as a function of the composition of many minerals and with the impurity concentration of semiconductors. Consequently this mode of SEM imaging is used in these fields of application. There are more problems associated with the cathodoluminescent mode than with most other imaging techniques in the SEM. Virtually all light detectors are also sensitive to electrons, so unless some technique is devised for rejecting electrons from the detector the signal generated by the light is overwhelmed by the signal generated by electrons. Although it is easy to reject low energy secondary electrons by biasing the detector negative as indicated in fig. 4.7, it is very much harder to prevent high energy reflected primary electrons from reaching the detector.

The final imaging mode which we shall cover in this section is probably going to be the most important in the future. The idea of forming an image from the electrons transmitted through a thin specimen in the SEM is simple and it can be achieved easily by moving the electron detector shown in fig. 4.7 from the side of the specimen to below it. At first sight there would seem to be little point in doing this since, as we have seen in the last chapter, transmission electron microscopes exist in a very highly developed form and are capable of far greater resolution than SEMs. However, a closer look at the scanning transmission technique (usually abbreviated to STEM) reveals some very great advantages over the conventional TEM method. These advantages arise principally because the SEM employs no lenses between the specimen and the image. The first benefit of this is that almost any electron which gets through the specimen can be detected and therefore can contribute to the image. To obtain a successful image in the TEM, on the other hand, we can only use electrons which have lost scarcely any energy while passing through the specimen; otherwise the range of electron wavelengths present in the beam would ensure that the image formed by the objective and projector lenses would suffer from gross chromatic aberration (see Section 1.6). Consequently, while we have to use an extremely thin specimen in the TEM, a SEM operating in the transmission mode (i.e. a STEM) can tolerate a specimen thickness perhaps three times greater while using incident electrons of the same energy. This has three beneficial effects: specimens are easier to prepare, larger areas of each specimen can often be viewed and the structures seen in thicker specimens are more likely to represent those present in the bulk material.

A second great advantage of the STEM is that the image, being merely a fluctuating current in a wire, can be manipulated in all the ways described in Section 4.5. The quality of image which it is possible to record should therefore be much higher than can be achieved on a photographic plate in the TEM.

It is a fortunate quirk of nature, beyond the scope of this book to explain, that the images of thin specimens seen by a STEM are virtually identical to those formed in a conventional TEM. All the theory of scattering contrast and diffraction contrast developed in Section 3.2 can be applied to STEM images, so any of the illustrations used in Chapter 3 would serve as demonstrations of the potential of STEM, including even the diffraction patterns (see next section).

One limitation of STEM techniques is very obvious if an ordinary SEM is being used. The resolution must be limited to the electron beam diameter, which can rarely be made smaller than 5 nm in a conventional SEM using a thermionic electron gun. It is necessary to use the much brighter field emission electron gun described in Section 2.3 if a smaller beam diameter, and hence better resolution, is to be achieved. However, so great has been the recent advance in field emission technology that STEM microscopes are now in operation using beam diameters of as little as 0·3 nm. While at present this performance is extremely difficult and expensive to achieve it opens up a whole realm of possibilities for the study of the arrangements of individual atoms. As a demonstration of the power of the technique pictures have been taken of rows of thorium atoms in a complex molecule, as illustrated in fig. 4.11. There is therefore nothing that a TEM can do which a STEM instrument cannot improve on, although at the moment it would be rather expensive to do so.

4.4. *Diffraction information in the SEM*

In the early years of commercial scanning electron microscopy (1965–70) the area to which the SEM could not usefully contribute was crystallography. The study of the crystal structures and relative orientations of different parts of the specimen, which could be achieved so readily by electron diffraction in the TEM, was apparently beyond the reach of the SEM. However, more recently it has been realized that it is possible to use the SEM as a tool to investigate the crystallography of both solid and thin specimens.

The first hint of information about crystal structure appearing in a SEM image came when large perfect crystals of silicon were being studied at low magnification (20 to 50 ×). The expected featureless images from the supposedly flat specimen were seen to be criss-crossed with pairs of straight lines. An example is shown in fig. 4.12. It was soon realized that these patterns contain almost the same

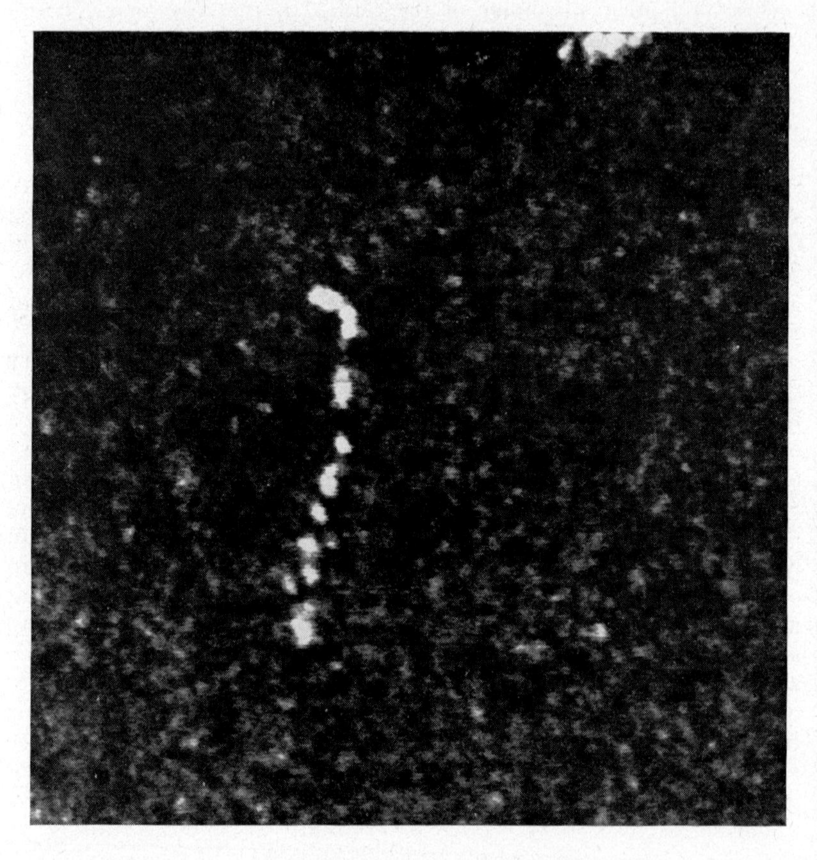

Fig. 4.11. A row of single thorium atoms observed with the scanning transmission electron microscope. ×1 600 000. (Courtesy of Professor A. V. Crewe, Enrico Fermi Institute, University of Chicago.)

information about the crystal structure of the specimen, and its orientation, as do conventional electron diffraction patterns, and in particular they resemble Kikuchi patterns (Section 3.3). The patterns are nowadays known as *electron channelling patterns* (ECPs) because of their mechanism of formation. It is clear from fig. 4.6 that as the electron beam is scanned across the specimen the angle, θ, at which it is incident on the specimen surface changes through a range of values. At low magnifications, where the side length of the raster may be 5 mm and the working distance (W.D.) in the region of 10 mm, θ will vary within a range of 50°; as the magnification is increased so this range of angles becomes smaller. If the specimen is a single

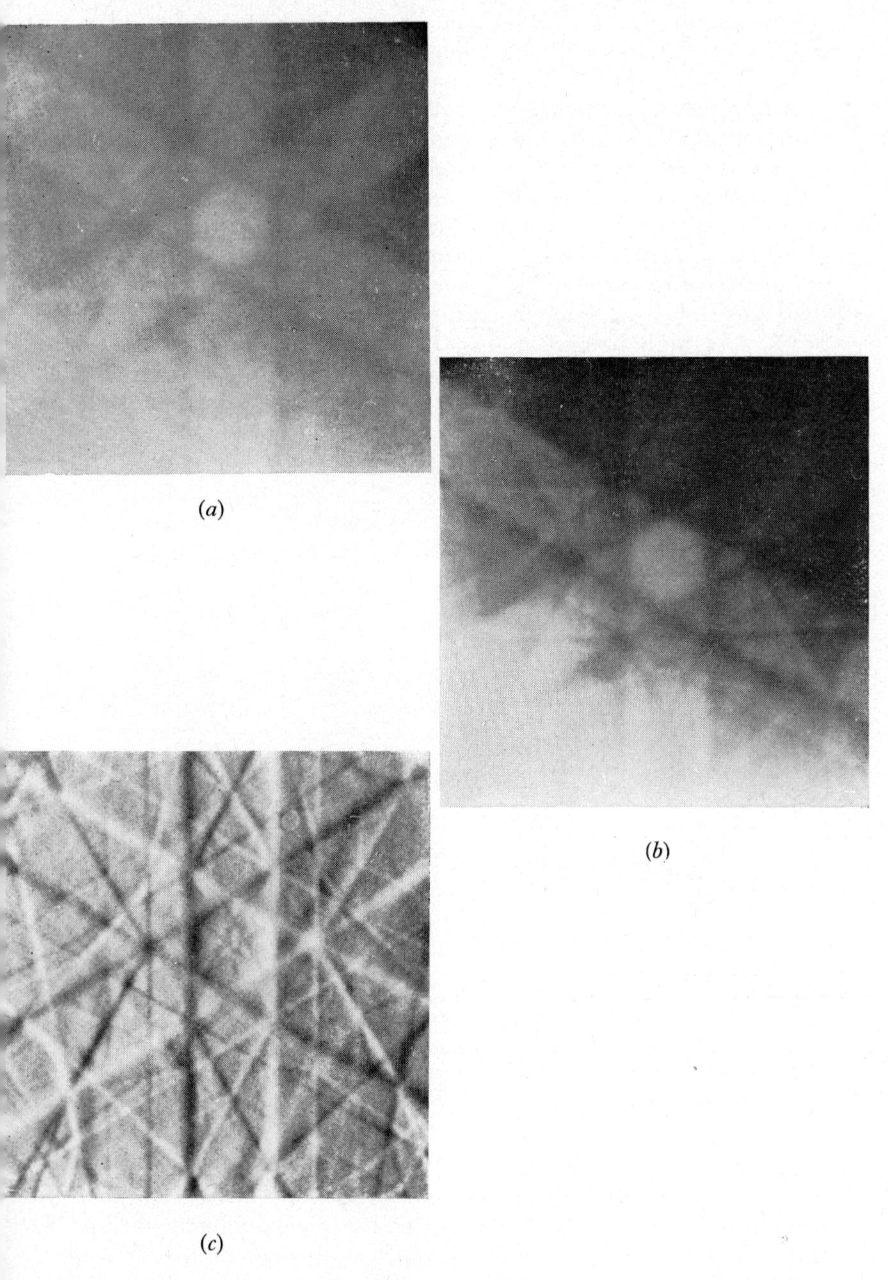

Fig. 4.12. An electron channelling pattern (ECP) from a crystal of silicon. (*a*) Viewed in secondary mode; (*b*) with added contrast; (*c*) differentiated.

crystal for the whole area scanned by the beam, then quite often the beam must be at the Bragg angle to sets of lattice planes. Figure 4.13 (*a*) shows the idealized situation for one set of planes: at two points in each scan (shown as full lines) the electron beam hits the same set of planes at its Bragg angle (see equation (2.9)). The effect of this is that at these two points the incident beam is strongly diffracted and penetrates farther into the specimen (i.e. is channelled). Consequently fewer secondary electrons or backscattered electrons are generated from these points. Since the same happens on each line of the raster the image shows a pair of dark lines corresponding to each set of lattice planes which lie within the range of angles scanned by the beam. At high magnifications, therefore, very few planes can be imaged and the ECP shows very few lines—the lower the magnification the more lines which are visible and the easier the pattern is to analyse.

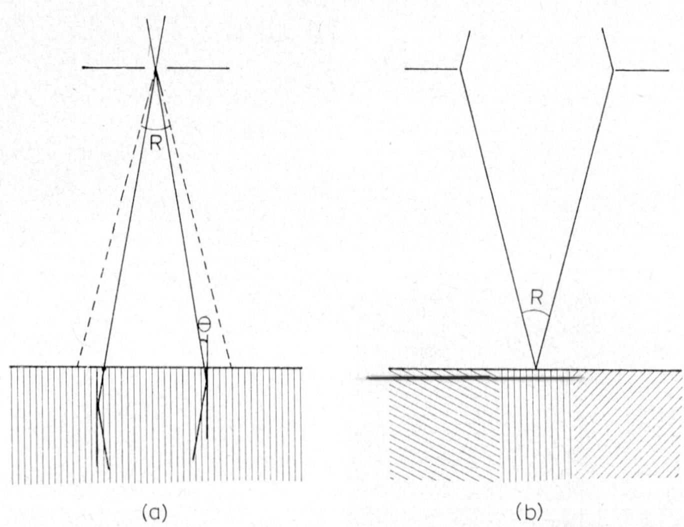

Fig. 4.13. Techniques for the observation of electron channelling patterns. (*a*) Using the same scanning of the beam as is used for normal imaging an ECP can be generated from a large region of single crystals; (*b*) if the beam is rocked about a point on the specimen surface an ECP can be generated from a small selected feature. R = range of angles scanned.

If the geometry of fig. 4.13 (*a*) is carried to its logical conclusion we find that the spacing of the pairs of lines in an ECP is inversely proportional to the lattice spacing of the planes which give rise to them. Thus this spacing is the exact equivalent of the spacing of spots in the conventional electron diffraction pattern; similarly the angles between pairs of lines are the same as the angles between spots.

Consequently an ECP contains just the same information as a transmission electron diffraction pattern.

A big disadvantage of the ECP technique as we have described it is that a large area of single crystal is needed for a useful pattern to be generated. This problem is partially overcome if the beam scanning system is slightly altered so that the beam rocks back and forth on the same spot, as shown in fig. 4.13 (*b*). The same range of θ values can then be scanned but all the ECP information is collected from a small area of the specimen. By this technique it is quite easy to form an ECP from an area about 5 μm in diameter and with care 1 μm ' resolution ' can be achieved. Clearly it is essential that a microscope can be used with both the scanning modes illustrated in fig. 4.13 so that an area may be imaged (using the normal scan of fig. 4.13 (*a*)) before a small part of it is selected for study by ECP (fig. 4.13 (*b*)).

The second method of using the SEM to obtain information about the crystal structure and orientation of the specimen arises when a thin specimen is examined by the scanning transmission (STEM) technique discussed in the previous section. It is very simple to see how diffraction patterns are formed in a STEM instrument if we invoke the principle of reciprocity, which Helmholtz propounded in 1886. This principle implies that any optical system works identically whether the rays are considered to be travelling in one direction or the other. Using this principle we can see that we can make the optical system of a STEM exactly equivalent to that of a conventional TEM. Figure 4.14 illustrates this point: the schematic optical system for a TEM operating in the diffraction mode is shown in fig. 4.14 (*a*) (compare with fig. 2.10) while the identical system with the rays pointing in the opposite direction (fig. 4.14 (*b*)) refers to the STEM instrument. We can recognize this more easily if we turn the STEM round (fig. 4.14 (*c*)) so that it is drawn in the conventional manner with the electrons travelling downwards. The only difference between the ray diagrams is the labelling of the various components—we can see that the electron source (gun) of the TEM is equivalent to the detector of the STEM and vice versa. Notice that we must put an aperture between the specimen and the detector in the STEM, corresponding to the condenser (illumination) aperture in the TEM, and that the beam is ' rocked ' on the specimen (as it was for the generation of ECP's) rather than being scanned across a raster.

The exact equivalence of the TEM and STEM optical systems means that the same electron diffraction information can be displayed in both cases. However, because of the electrical nature of the signal in the STEM electron intensities in the diffraction pattern are more easily quantified, enabling more information about the degree of perfection of the crystal specimen to be deduced—yet again a case of " anything TEM can do, STEM can do better ".

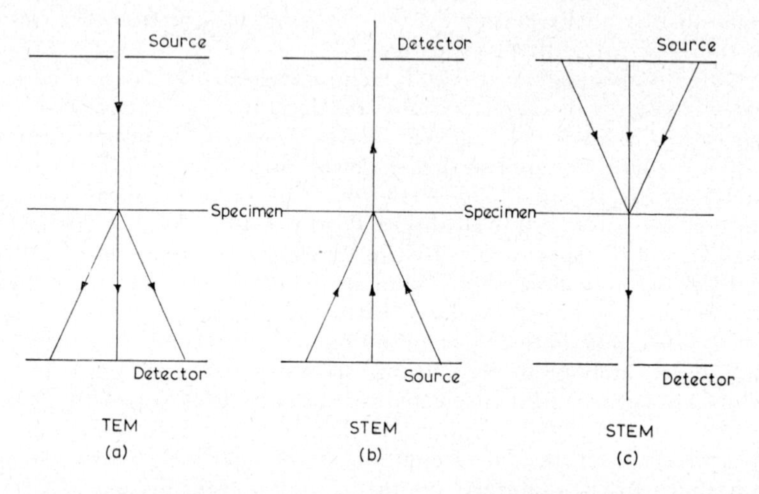

Fig. 4.14. The simplified optical systems of the transmission (*a*) and scanning transmission (*b*) microscopes. In (*c*) the STEM system is inverted from (*b*) so that it is ' the right way up ' with the electron source at the top and the detector at the bottom, as in the conventional transmission microscope (*a*). (*a*) and (*b*) are clearly identical except for their nomenclature.

4.5. *The presentation and recording of images*

The principle by which information is displayed in a scanning electron microscope is almost always the same. The chosen secondary effect—let us say secondary electrons, for example—is detected by a suitable detector and the magnitude of the resultant electrical signal is used to alter the brightness of the spot on a cathode-ray tube. In practice the appearance of the image depends on the number of frames which the c.r.t. spot (and therefore the electron beam on the specimen) is allowed to scan per second. If the scan is carried out at the rate of 50 frames per second the display appears simply as a television image. This is very convenient. If the specimen is moved, or the image needs to be focused, the changed picture appears instantly on the screen. We can even examine moving mechanisms in the microscope—a small watch makes a fascinating object to study. However, it is not always possible to scan the specimen at the fast rates needed for a television display and often we must be content with a frame being scanned every one or two seconds. At these rates our eye can see the horizontal line moving down the screen, and we must use a c.r.t. having a long persistence phosphor; such a screen continues to emit light for some time after the c.r.t. spot has passed and therefore the image persists for a couple of seconds until the spot returns on its next scan of the frame. The effect of this slow scan rate is that the operator of the SEM sees an image which always has a bright line moving down it. This is initially confusing but most people rapidly become accustomed to it.

Why should it be necessary to scan the specimen at such a slow rate? The reason is that the electron beam must be allowed to spend sufficient time on each point of its raster to generate a signal which is big enough to be measured accurately. If the secondary electron mode is in use this ' signal ' is the current generated in the electron detector by the arrival of a discrete number of secondary electrons. If the beam only lingers long enough on each point of the raster for two or three electrons to be given off, then there will be a great deal of uncertainty in the signal. In other words, an error of one or two electrons alters the signal very drastically. Since secondary electrons are emitted at random intervals, not at a constant rate, simple statistical theory tells us that we can only specify that N electrons arrived in a certain time with a certainty of $\pm\sqrt{N}$ and therefore the likely percentage error in any measurement is $\sqrt{N}/N \times 100 = 100/\sqrt{N}$. Clearly two or three secondary electrons per point will not provide an accurate enough signal—the only ways of increasing the signal from each point are to increase the current in the electron beam (i.e. the number of incident electrons per second) or to slow down the scan (i.e. linger longer on each point). At low magnification and resolution the former course is acceptable and therefore television rate displays can be used. However, if high resolution images are required at high magnification it is not possible to increase the current in the electron beam without increasing its diameter and hence making the resolution worse. The only course available therefore is to slow down the scan.

The best quality images are needed for photographic recording, which is carried out with a camera focused on a special high quality c.r.t. The scan rate is normally slowed right down so that one frame takes 20, 40 or even 100 s, while the camera shutter is open. This provides a good quality permanent record of a static image. If dynamic effects are occurring in the microscope, for instance the movement of the gears in a working watch or the fracture of a metal specimen, these are most easily recorded from a television display. In this case it is not necessary to photograph the screen with a cine camera since the television signal can be led directly into a video tape recorder. The limitations of quality in a video system must then be added to the limitations inherent in the television rate scanning so that high resolution images cannot practically be recorded in this way.

In earlier sections it has been stressed that one of the big advantages of the SEM is that the image information is available as an electrical signal and can therefore be processed before it is displayed. There are really two ways in which we can manipulate the appearance of the image of any specimen before we choose to photograph it, which involve the two major components of the image-forming system, the detector and the signal amplifier, shown very schematically in fig. 4.15. The most obvious way in which we can influence the appearance of the specimen is in our choice of detector. We can select the secondary effect to be displayed (i.e. the mode of operation of the SEM) by taking the signal from one of several sources, perhaps an electron detector, an X-ray detector, a light detector or the specimen current. The choice of detector determines what information we are going to see, and, as fig. 4.8 has already shown, a small change in the detector can drastically alter the image. However, even when the detector has been selected (and most SEM micrographs are taken using the secondary electron detector) there are many ways in which the same information can be displayed, using differences in the signal processing stage shown in fig. 4.15. To take an easy example: the main function of the signal processor is to amplify a small signal, *i*, into a bigger signal, *I*, which controls the brightness of the c.r.t. display. Thus by controlling the *gain* of the amplifier we can control the magnitude of *I* and hence the overall brightness of the image.

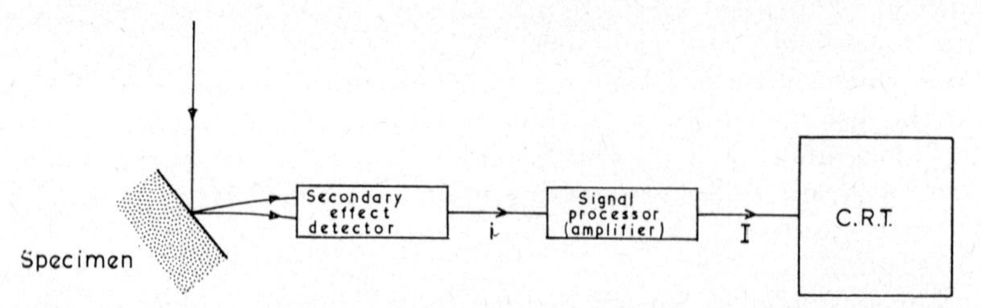

Fig. 4.15. The main components of the image forming system in a scanning electron microscope.

It is helpful in discussing even such simple signal processing if we consider the variation in the signal across one line of the raster scan. We can then plot the signal magnitude (*I*, and hence the brightness of the c.r.t. screen) against distance across the picture. An example of this plot is shown schematically in fig. 4.16 (*a*); usually this information is available all the time on an SEM, being displayed on a separate small c.r.t. The effect of changing the signal amplifier gain, and therefore the image brightness, will be seen as a vertical shift of the whole plot. If the gain is turned up too much the brighter parts of the picture will all appear totally white and no detail will be visible in them. This effect can be seen in the micrograph of a glass fibre in fig. 4.16 (*b*). It is evident from micrographs (*a*) and (*b*) that it is not really possible to display both the fibre itself and its background so that the details of both can be easily seen. It is even difficult to set the brightness so that the edge and the centre of the fibre can both be seen clearly. There are a couple of fairly simple electronic ways of improving this situation: the first is to alter the action of the signal processor so that I is not linearly proportional to i—in other words, so that small signals are amplified more than large signals and less

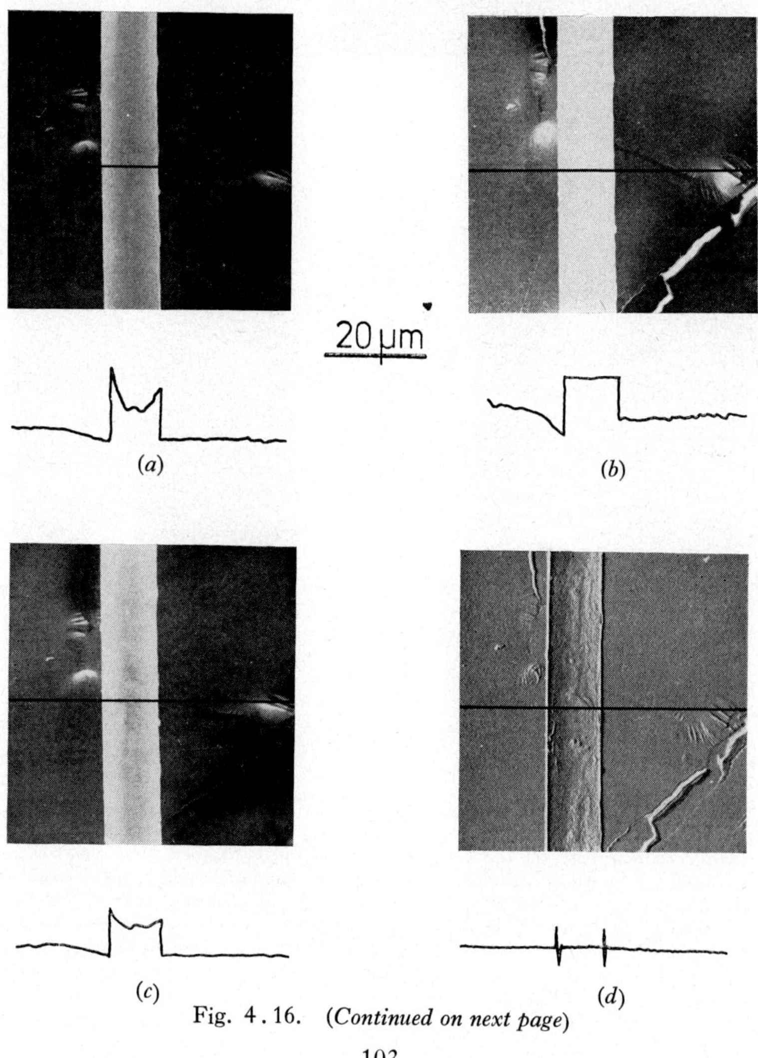

Fig. 4.16. (*Continued on next page*)

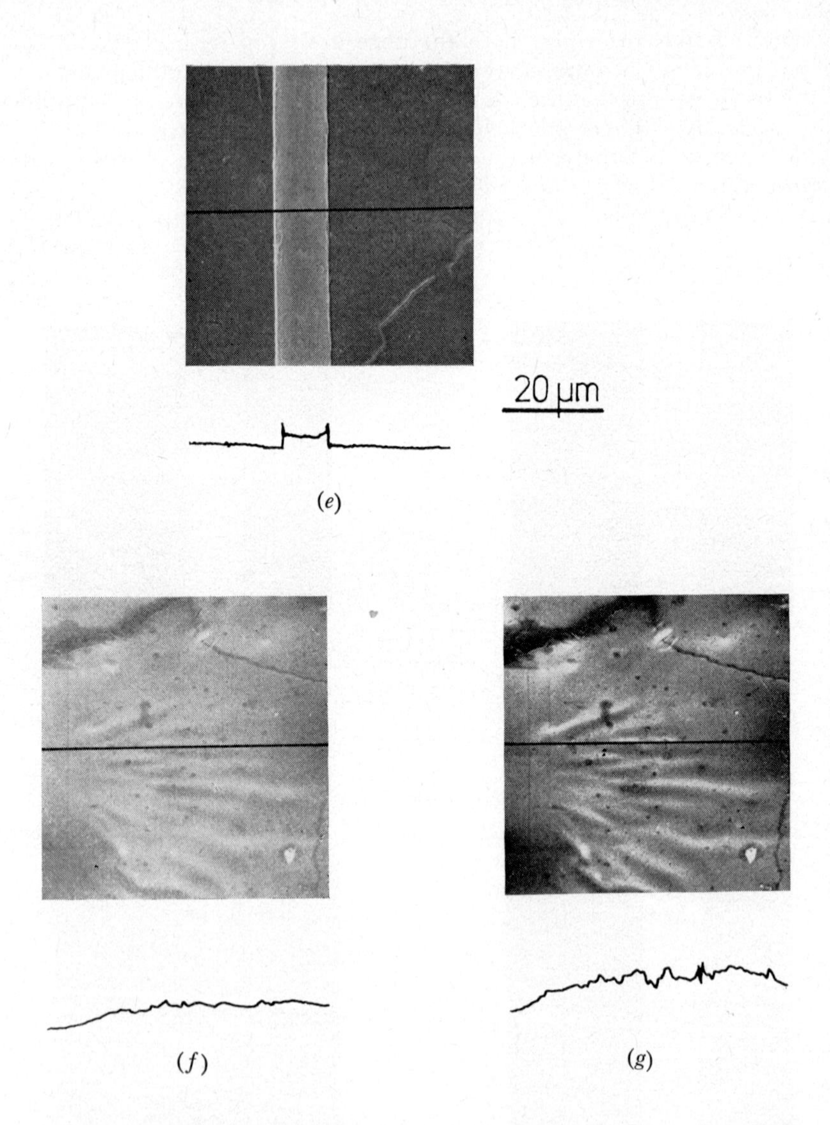

Fig. 4.16. Examples of signal processing in the SEM. (*a*) Normal secondary electron image of a glass fibre; (*b*) as (*a*) with increased brightness; (*c*) as (*a*) with contrast suppression (gamma increased); (*d*) as (*a*) with the signal differentiated; (*e*) a mixture of (*a*) and (*d*); (*f*) normal secondary image of the adhesive tape on which the fibres were mounted; (*g*) as (*f*) with contrast enhancement. In every case the intensity distribution shown is that along the horizontal line marked on the micrograph.

contrast is retained in the image. This is known as varying the 'gamma' of the system in analogy to the gamma value used to describe the sensitivity of photographic emulsions. The effect of altering gamma is shown in fig. 4.16 (*c*). The other approach is to electrically differentiate the signal; the effect on the signal is shown in fig. 4.16 (*d*). As the micrograph shows, the overall background is grey and only local variations in contrast can be seen. Consequently fine detail is visible in both the formerly bright fibre and the dark background. Unfortunately the differentiating process removes the illusion of a three-dimensional structure from the micrograph. A better compromise is often produced by mixing the original and differentiated signals, with the result shown in fig. 4.16 (*e*).

Many other signal processing methods are possible but the only other technique in everyday use is the artificial enhancement of contrast from 'flat' specimens. Figure 4.16 (*f*) shows a micrograph of the double-sided sellotape on which the glass fibre was mounted. This is a fairly flat object which does not show a great deal of contrast; its contrast can be enhanced by subtracting a constant signal from the whole plot and then amplifying the remaining signal more by turning up the gain ('brightness'). The effect is shown quite markedly in fig. 4.16 (*g*).

Of the methods of manipulating the signal which have been discussed in this section, three are available on virtually every SEM; these are the ability to slow the scan rate for photography, to control the brightness (amplifier gain) and to adjust the contrast. The other techniques mentioned, and many more, are used where the particular applications of the SEM justify their cost.

4.6. *The preparation of specimens for examination in the SEM*

Because in most cases the SEM is used to study surface morphology, bulk specimens are normally used and specimen preparation is far simpler than for transmission microscopy. Since there are no lenses below the specimen there is a great deal of space available to accommodate the specimen and various mechanical controls for moving it. For example, it is possible in some modern microscopes to mount a specimen as large as $150 \times 50 \times 50$ mm, which can be moved around within the vacuum system so that large portions of it can be viewed. This means that it is possible to mount many specimens without even cutting them up. However, there is one prerequisite for effective viewing: the surface of the specimen must be electrically conducting.

The necessity for a conducting specimen arises from the statistics of electron yield when any material is bombarded with electrons. We can express the number of electrons emitted from the surface per unit time as a current i_S. This will contain both 'secondary' electrons and 'reflected' primaries. In general this is not equal to the number of primary electrons incident on the surface per second, expressed as the current i_p. We define *electron yield* as the ratio i_s/i_p and find that for virtually all materials it varies with the energy of the primary beam in the manner shown in fig. 4.17. The implication of this curve is that there are only two operating voltages for the microscope where the yield is unity and hence electrons are leaving the surface at the same rate as they are hitting it. For most materials these two crossover voltages (V_{c_1} and V_{c_2}) are much lower than the operating voltage of the microscope. Consequently, during normal operation there is a surplus of electrons building up on the specimen surface. If these are not conducted away to earth the specimen surface will become negatively charged until very soon the incoming primary electrons are repelled and deviated from their normal path and a distorted image will be formed.

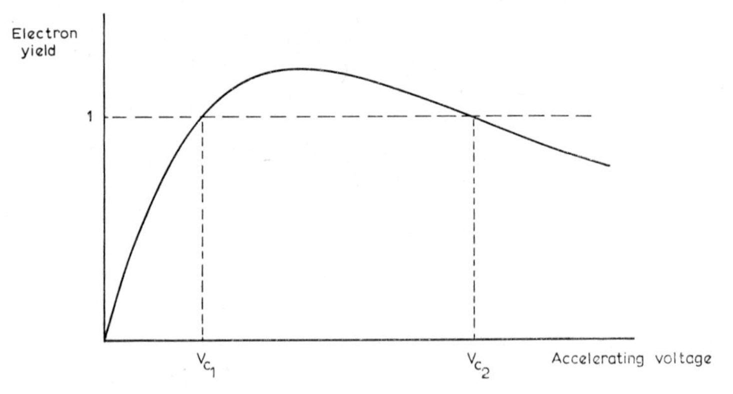

Fig. 4.17. Secondary electron yield (i.e. the number of secondary electrons emitted per incident primary electron) versus accelerating voltage. The two voltages at which the yield is unity are known as the first (V_{c_1}) and second (V_{c_2}) crossover potentials.

Clearly there is no difficulty in studying metal specimens. It is only necessary to ensure that there is not a layer of non-conducting material such as oil or grease on the surface of the specimen, and to fasten it in the microscope so that it is in electrical contact with earth. However, non-conductors such as ceramics, polymers and biological materials present more of a problem. There are two solutions: either the microscope must be operated at an electron accelerating voltage corresponding to the second crossover, V_{c_2}, or the specimen surface must be made to conduct electricity. Operation of an SEM at such low voltages as V_{c_2} is not often desirable, both because resolution is impaired and because there is no guarantee that V_{c_2} is the same for all parts of the specimen. Consequently the second approach, involving coating the specimen with a conductor, is usually employed.

It is quite a simple matter to evaporate a very thin layer of metal, perhaps only 10 nm thick, onto a solid specimen. Such a thin layer does not alter the detail visible in the SEM but is thick enough to conduct away the excess charge. The metal used most frequently is an alloy of gold and palladium, because it is a very efficient emitter of secondary electrons, and it is applied to the specimen by vacuum evaporation. This technique involves placing the specimen in a vacuum bell jar, in which a small amount of the gold alloy is melted. A typical arrangement is shown in fig. 4.18. The alloy is evaporated on to every surface inside the bell jar, including the specimen, in a few seconds. Thus any specimen which can tolerate the vacuum

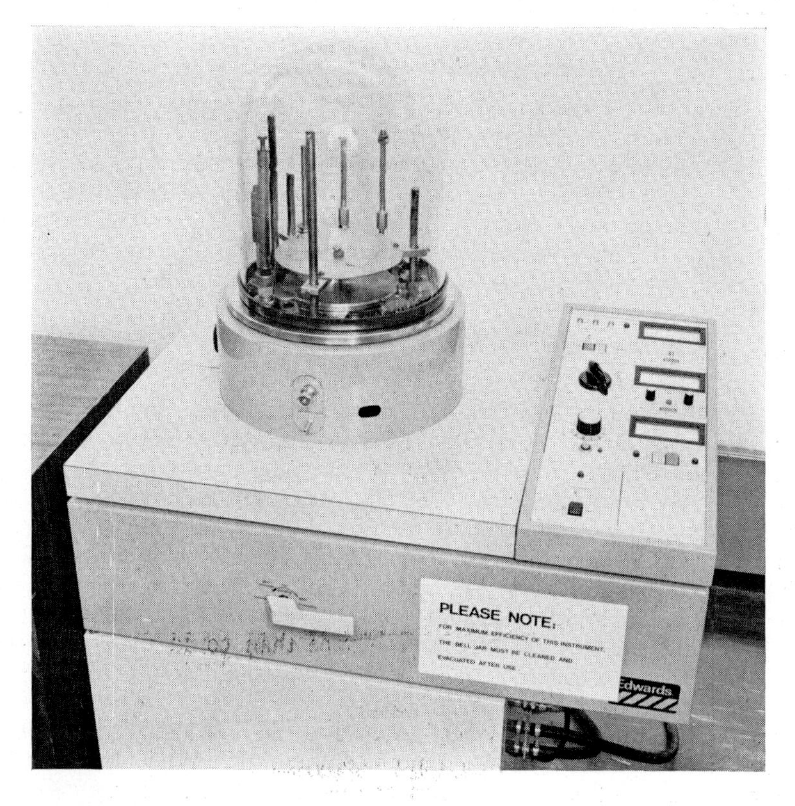

Fig. 4.18. A vacuum coating unit. Inside the bell jar a thin layer of carbon or metal can be evaporated and deposited on to the specimen.

without damage can be coated in this way. Since the specimen must anyway be placed in a similar vacuum to be examined in the SEM this is hardly a restriction, although it does emphasize the need to be sure that evacuation does not alter the appearance of the specimen, for instance, by causing a biological specimen to collapse. The vacuum coating procedure does take about half an hour, however, and if it is necessary to examine a specimen without delay at fairly low magnification (say 5000 × or less) an alternative approach can be adopted. Organic conducting materials are available (for use in the textile industry as antistatic agents) in an aerosol form and can be sprayed lightly on to the specimen, which can then be inserted in the microscope within a minute or two.

Clearly this rapid preparation is a far cry from the highly skilled, time-consuming preparation of thin specimens for TEM work. This is one of the factors which account for the ever-increasing popularity of the SEM technique at all levels of science, technology and industry.

4.7. *Examples of the application of scanning electron microscopy*

The range of problems to which SEM has been applied is probably wider than for any other type of electron microscope. In almost every field of scientific and industrial research and development it has been found helpful to be able to study fine details at high magnification and large depth of field. The applications which are described in this section are but a few of the many which have been undertaken in just one laboratory over the past few years.

4.7.1. *Identification of diatoms*

The diatom is a fascinating subject for study by SEM. There is such a large number of types, each of which may differ in only a small way from the other. Because of their pronounced three-dimensional nature and their small size, diatoms have long been used as objects with which to test the performance of a light microscope. The piecing together of the three-dimensional shape of these organisms is a very time-consuming job if only a light microscope is available. The advent of the scanning microscope has meant that a single SEM photograph can yield more information both about the shape of the diatom and the fine detail of its surface than could many a light micrograph, even when taken by a superbly skilled microscopist. An example is shown in fig. 4.19; this diatom, a *Cyclotella meneghiniana*, was found in the River Wey near Guildford, England and appears to be an unusual variant of the much more common ‘ Christmas cake ’ type shown in fig. 4.20. Discoveries of this type are much more likely using the SEM, since literally thousands of diatoms can be mounted in the microscope at one time and all of them can be studied either

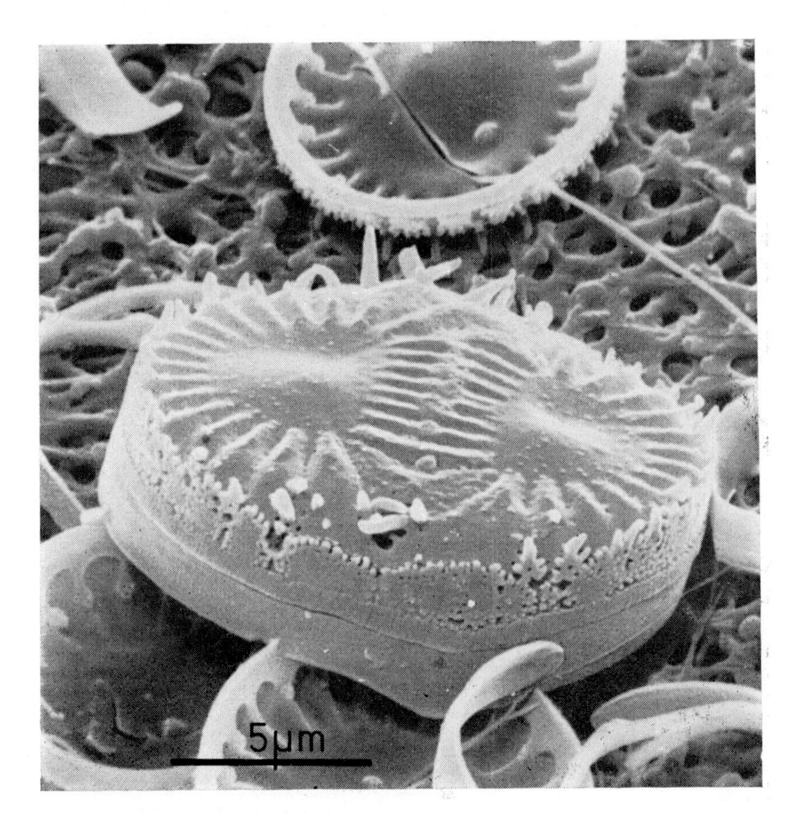

Fig. 4.19. An unusual 'double' variant of the diatom shown in fig. 4.20. (G. Gibbs and M. O. Moss.)

very rapidly or in great detail. It is possible to scan quickly through several thousand diatoms in a morning just to see if there are any unusual types worthy of further study.

4.7.2. *Study of a bedmite*

This application is not truly scientific but it does illustrate the graphic power of an SEM micrograph. The bedmite (*Dermatophagoides farinae*) shown in fig. 4.21 is a type commonly found in bed-linen and mattresses in even the cleanest homes! It passes unnoticed because it is only 0·4 mm long, and for the same reason is not easy to study by any other technique. The big problem is to mount the mite in the microscope without knocking off its antennae or the hairs on its legs. Many of these insects were mounted

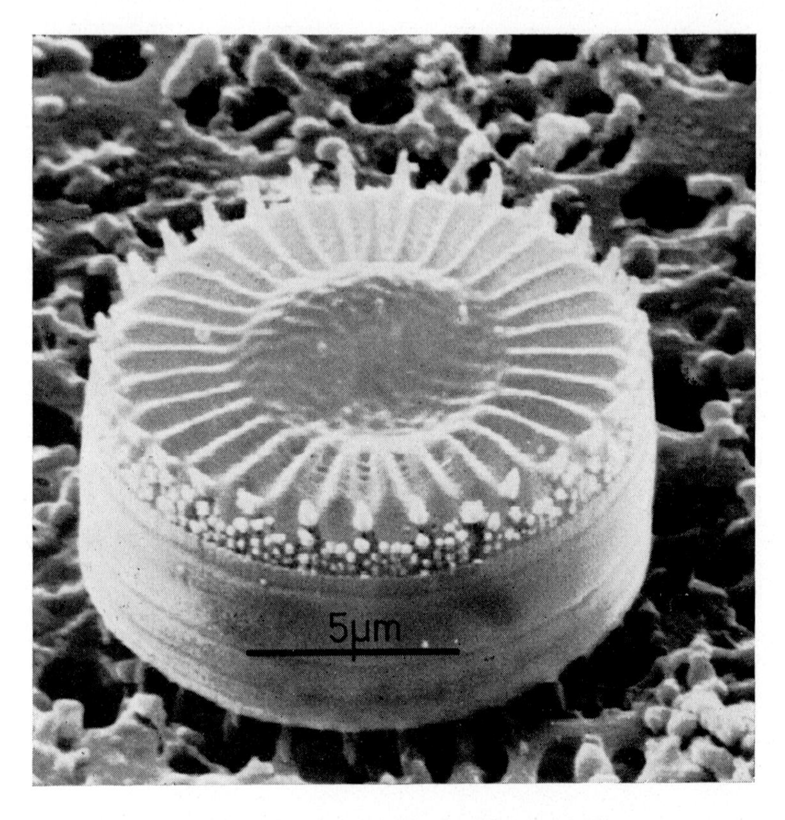

Fig. 4.20. The diatom *Cyclotella meneghiniana*.
(G. Gibbs and M. O. Moss.)

unsuccessfully before this rather lifelike specimen was photographed. It is fascinating to compare the result obtainable in a couple of minutes with an SEM with the famous drawing by Robert Hooke (published in 1665) of a flea (fig. 4.22). The original of this astonishing drawing is over 40 cm long and must have represented days or even weeks of painstaking work with a very crude compound microscope, to say nothing of the skill required in the sketching.

4.7.3. *Fracture of a composite material*

Fibre composite materials are becoming more and more widely used in everyday engineering applications. By ' fibre composite ' is meant a material in which a fibre of one material (glass, carbon, asbestos, steel wire, etc.) is embedded in a matrix of another (nylon, epoxy resin, concrete, etc.). One composite material, fibreglass, has already become very familiar. In general these materials combine the high strength of the fibre with the greater flexibility or corrosion resistance of the matrix and the two together are often lighter than a conventional material, for instance an alloy, would be. Figure 4.23 shows a test fracture in an experimental composite material containing carbon fibres in an epoxy resin matrix. This forms an extremely strong, stiff, yet light material which is of considerable use where weight is critical, for example, in aircraft or spacecraft.

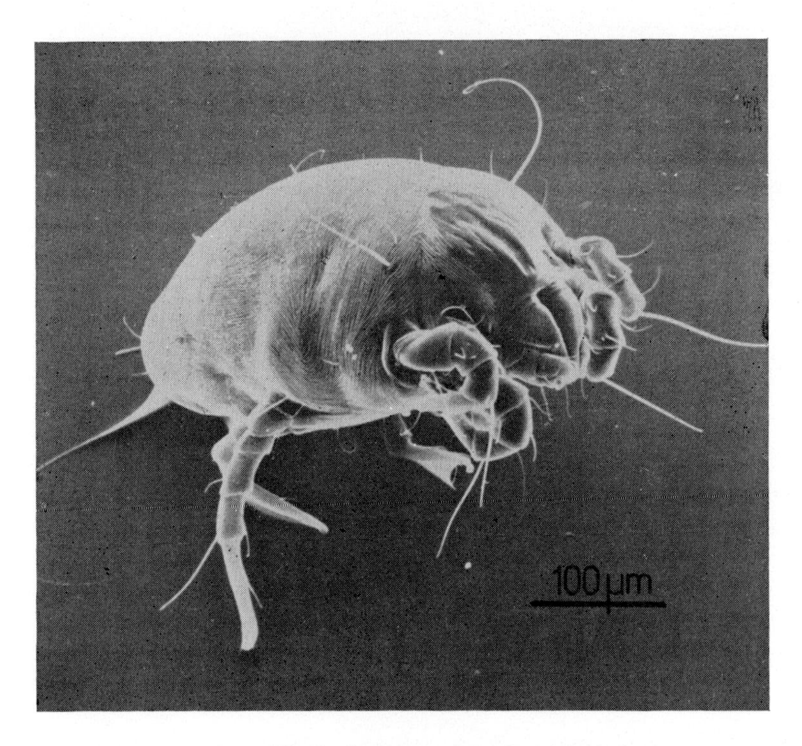

Fig. 4.21. The bedmite *Dermatophagoides farinae*.
(G. Gibbs and A. C. McDonald.)

However, composite materials fail in rather a different way from the more usual metallic materials when overloaded. The test piece shown in the figure has been broken by bending: half the fracture face appears smooth while the other half shows many fibres sticking out of the matrix. The reason is that first one side of the ' beam ' has failed under compression and then the opposite side has failed under tension. Fibres have been pulled out of their matrix on the tension side but not on the compression side. Scanning electron micrographs such as fig. 4.23 allow us to determine the relative sizes of the two zones and to measure the lengths of fibre which have been pulled out in the tensile region. This information is of tremendous use in enabling materials scientists to predict how, why and under what loads composite materials will break.

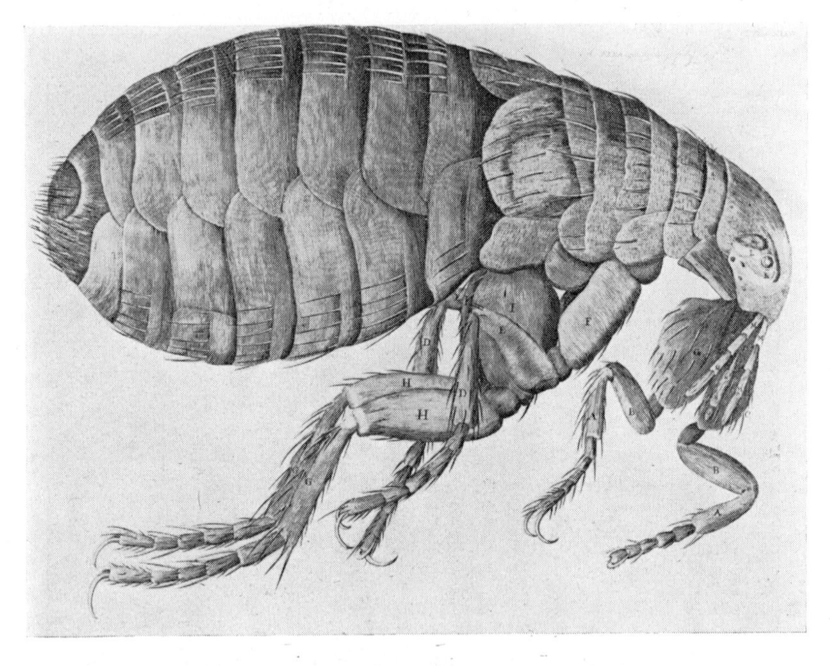

[*reproduced by kind permission of the Royal Institutson*]

Fig. 4.22. Robert Hooke's drawing of a flea, published in 1665, which is based on his observations with an early light microscope.

4.7.4. *Corrosion in power station components*

One of the big problems in power-station technology is corrosion of components which may be in contact with high-pressure steam or sea-water. One such problem arises in condenser tubes. In some tubes, made from an alloy similar to brass but containing a few per cent of aluminium, tiny pinholes have been found to occur. Obviously any small hole is potentially disastrous, since it may eventually go right through the tube. The puzzling aspect of this form of attack is why the corrosion should only occur at particular points and not over the whole surface of the tube. Figure 4.24 is a picture of one ' pinhole ', which is 0·1 mm in diameter—in other words, only just

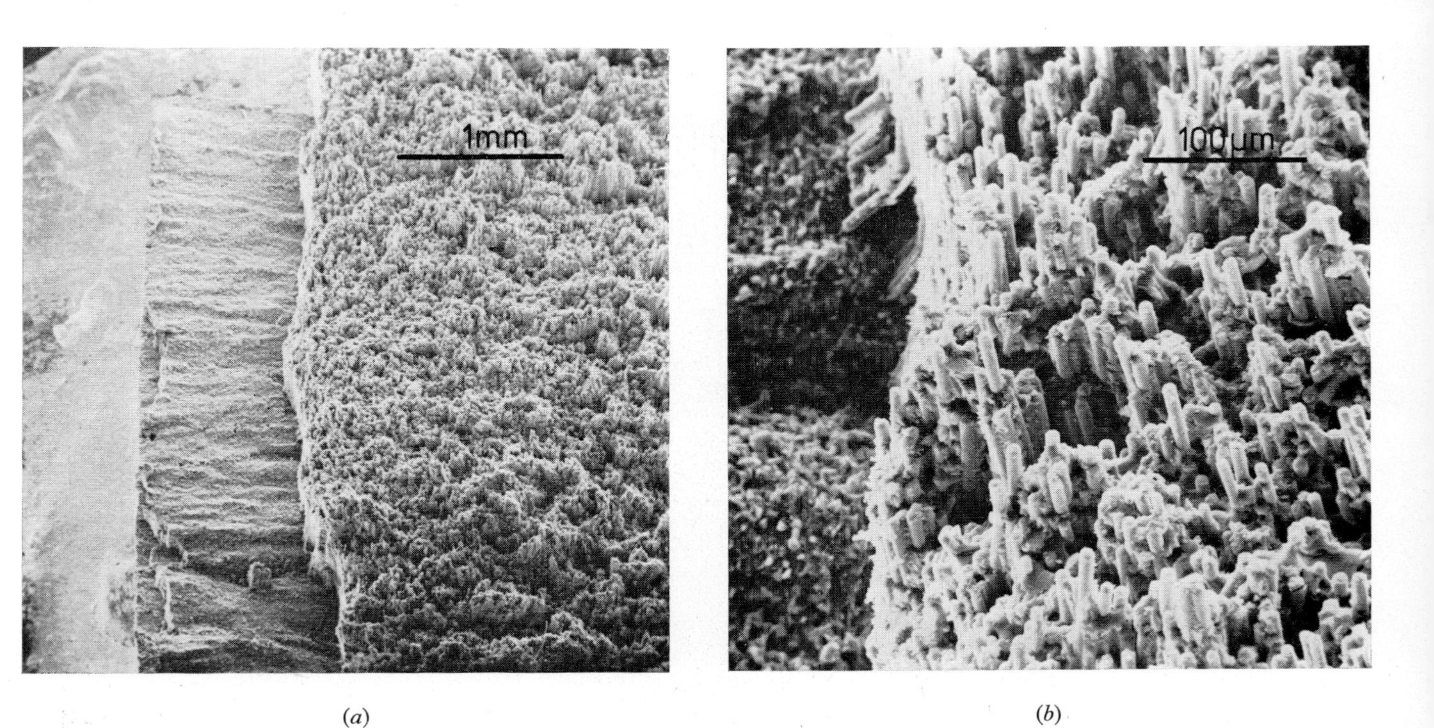

(*a*) (*b*)

Fig. 4.23. A fracture test specimen of carbon fibre/epoxy resin composite material. The low magnification micrograph (*a*) shows clearly the two different fracture zones, while at slightly higher magnification (*b*) the individual fibres can be seen.

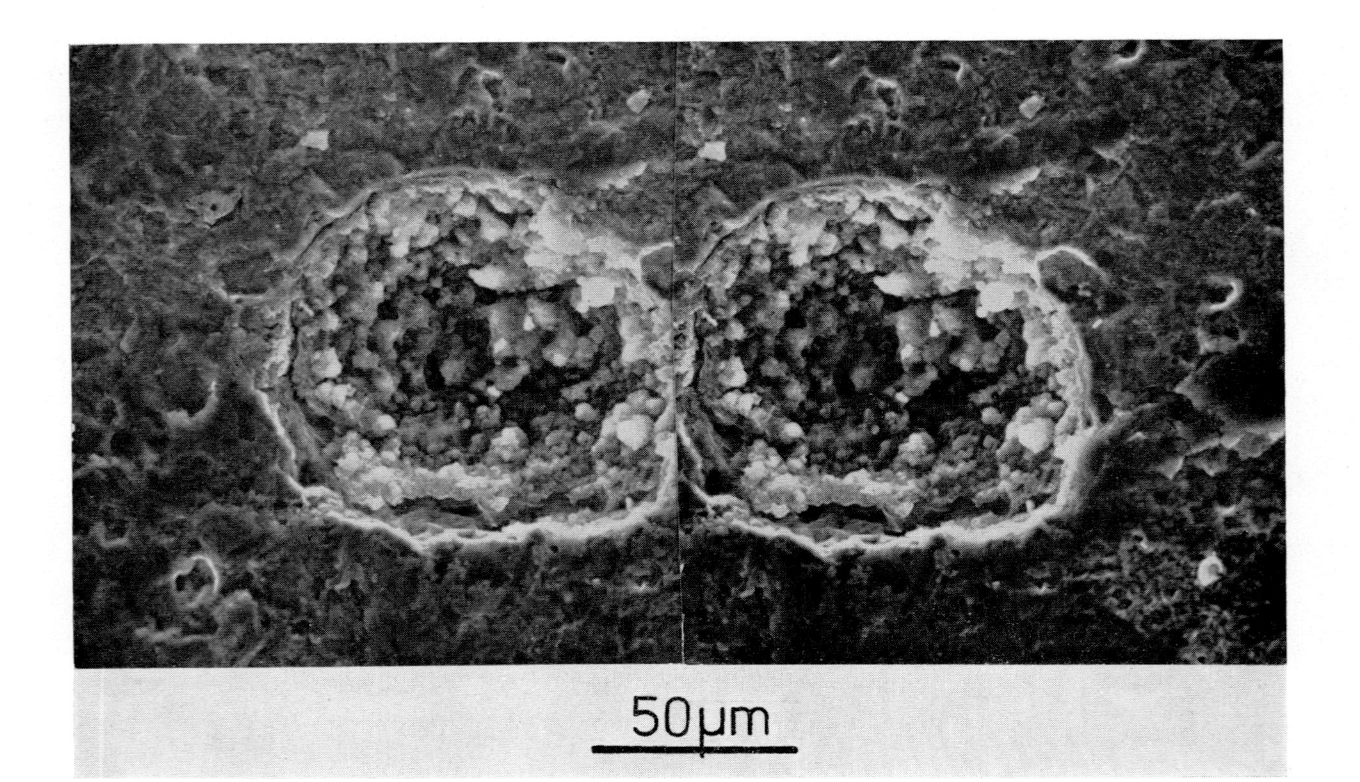

Fig. 4.24. A ‘ stereo pair ’ micrograph of a corrosion pit in a power-station condenser tube. The three-dimensional effect can be observed by viewing the pair of micrographs using a special viewer or, with practice, by the technique described in the text. (G. Gibbs and J. E. Castle.)

visible to the naked eye. The micrograph is presented as a ' stereo pair ' to illustrate how useful this technique is. The two pictures have been taken in identical conditions except that between exposures the specimen has been tilted a few degrees. The two micrographs therefore represent the views which our left and right eyes would see. If we look at the pair of pictures so that the left eye sees the left image and the right eye sees the right image our brain will present the combined images as a single three-dimensional picture. The easiest way of achieving this is to view the pair of images using a special pair of lenses on a stand, which hold the eyes in the correct place for stereo viewing. However, with a little practice most people who have fairly well-matched eyes can train themselves to get the same effect without any special viewing device. The stereo pair is held about 25 cm from the eyes (the closest distance at which it is comfortable to focus on it) and then the eyes are allowed to relax so that they are focused on infinity. Three very blurred images should now be visible. While mentally concentrating on the centre image, wait: slowly the eyes will come back to focus on the image and it will be seen in three apparent dimensions. Four or five attempts at this technique usually lead to success. Do not go on too long—ten or twenty attempts usually lead to a headache!

Once this micrograph is viewed in stereo the object of the exercise becomes apparent. Not only is the great depth of the pit evident, but it can be seen that the walls of the hole are not the solid metal they appeared in a single image but a veritable honeycomb of partially corroded alloy. Note too that the two white spherical particles to the upper right of the pit are really jutting out into space—they are scarcely noticeable in the single image. This sort of view tells us a great deal about the nature of the attack and the extent of the corrosion damage, although, to be fair, it did not absolutely provide the solution to the problem. A combination of these SEM observations with some electrochemical theory did however finally enable the cause to be traced.

4.7.5. *Structure of electrodeposited copper*

Electroplating is a widely used method of depositing a thin protective coating on to a cheaper metal base. A common example is the chromium plating of automobile parts. Much of the research and development of plating science is carried out using copper plating, because the solutions used are simple and relatively safe. The plated copper deposit tends to grow in crystals (grains) which are in the same crystal orientation as the base metal (the substrate). Consequently if copper is plated on to a copper base which is a single crystal (or nearly so) all the grains align. This effect can be seen in fig. 4.25; we do not need diffraction information to determine the orientation of the individual crystallites because their obvious facets disclose the underlying crystal structure.

The copper in this example was plated from an acid sulphate bath (a mixture of copper sulphate solution and sulphuric acid) but the size and shape of the grains, and hence the hardness and corrosion resistance of the deposit, is determined by various plating bath parameters such as the impurity content, the temperature or the current density. The effect of each of these parameters on the structure of the deposit can be seen in the SEM as a change in the type of faceting or a change in the average grain size. Since the copper electrodeposits require no specimen preparation the SEM provides a method of studying electroplating which is both graphic and quick.

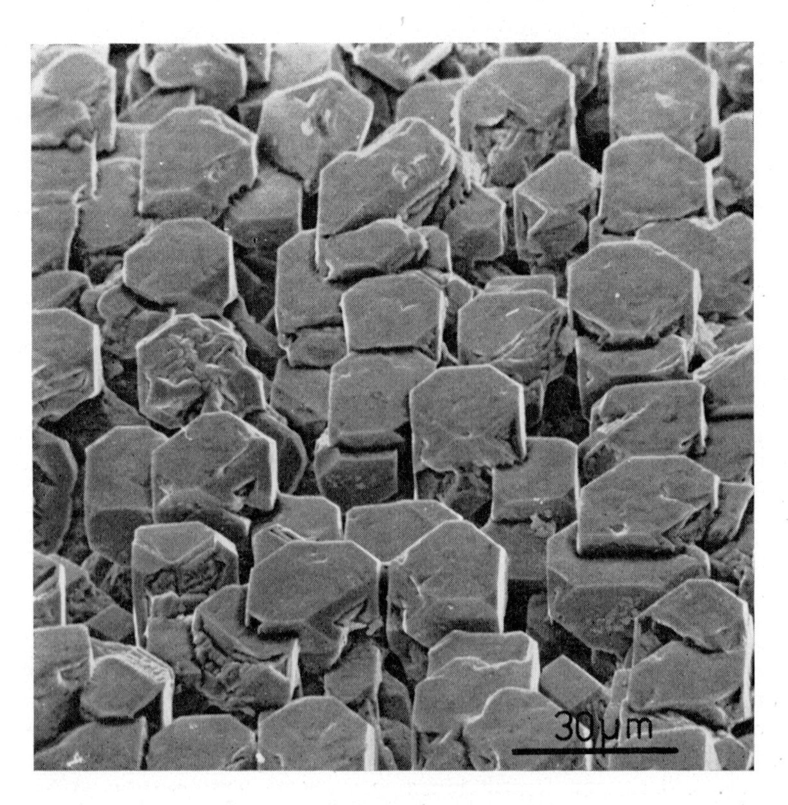

Fig. 4.25. The regular array of faceted grains on the surface of electroplated copper. (G. Gibbs and C. T. Walker.)

CHAPTER 5

analytical information from an electron microscope

It has been stressed in Chapter 2 that whenever electrons with several kilovolts of energy strike a solid specimen X-rays characteristic of the atoms present in the specimen are produced. In discussing the formation of images in the TEM and the SEM we have ignored these X-rays. However, to do so is to discard a great deal of information about the composition of the specimen. This was realized in the 1950s and since then efforts have been made to use all types of electron microscope as *microanalysers*. This term is not meant to imply that very accurate analyses are possible, but that an analysis can be performed on a very small amount of material, or, more usually, on a very small part of a larger specimen. This is just what the more conventional forms of analysis (chemical or spectrographic) cannot do and hence microanalysis in the electron microscope has become an important tool for both biologists and materials scientists.

In principle we can determine two things from the X-ray spectrum emitted by any specimen. Measurement of the wavelength (or energy) of each characteristic X-ray that is emitted enables us to find which elements are present in the specimens; measurement of how many X-rays of any type are emitted per second should also tell us how much of the element is present, i.e. its concentration. These two types of analysis, qualitative and quantitative, are equally important to the microanalyst, but may be carried out in different ways, as this chapter will show.

5.1. *The generation of X-rays within a specimen*

We have seen in Section 2.6 that bombardment of a material with high energy electrons will result in the emission of ' characteristic ' X-rays, whose wavelengths depend on the nature of the atoms in the specimen, together with white radiation of all wavelengths down to a minimum corresponding to the incident electron energy (fig. 2.8). (The reader to whom these concepts are unfamiliar might find it helpful to re-read Section 2.6 before continuing.) Before we can use these X-rays for analytical purposes we need to know which of the many characteristic X-ray lines for each element is the most intense; this enables us to choose the best line to use as an index of how much of each element is present in the sample. The situation at first appears to be very complex since, as fig. 5.1 shows, there is a large number of electron transitions possible in a large atom, each of which should lead to an X-ray of a unique wavelength. It transpires, fortunately for the microanalyst, that in the K series the lines $K_{\alpha_1}+K_{\alpha_2}$ (which may be so close together that they can rarely be distinguished) are seven to eight times more intense than $K_{\beta_1}+K_{\beta_3}$ (another close pair). Consequently the K_α ' doublet ', as it is called, is most frequently used for analysis. However, it is not always possible to excite the K series of lines in an electron beam instrument since, as the atomic number of the emitting element increases, the energy required to knock out a K-shell electron also increases. For example, elements heavier than tin ($Z=50$) need electrons of more than 25 keV to excite any K lines at all, and are not efficient producers of K X-rays until the incident electron energy is about 75 keV. Since we would like to be able to analyse a specimen in an SEM, where electron energies of only perhaps 30 keV are available, we must

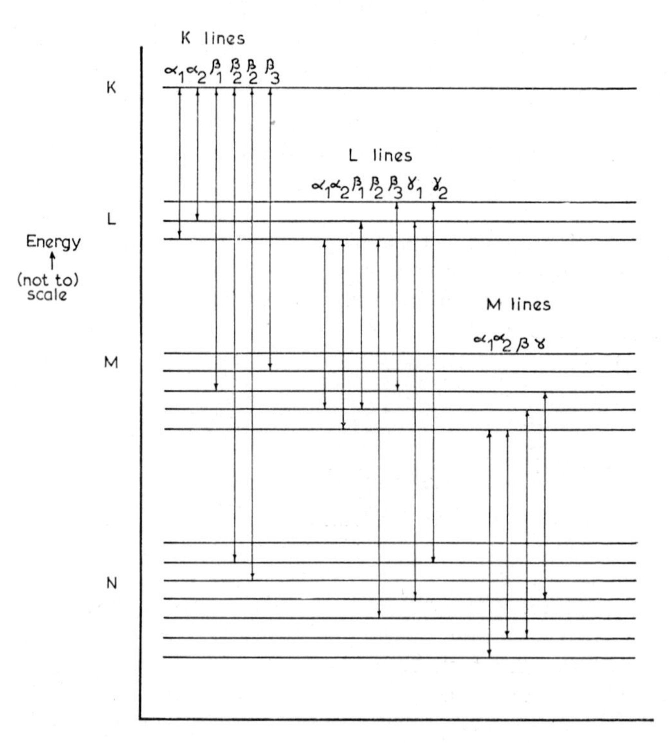

Fig. 5.1. Some of the more common transitions between the K, L, M and N shells of an atom which lead to the X-ray lines indicated. The nomenclature is that of fig. 2.1.

look for other characteristic X-rays which are more easily excited in order to detect heavy elements. Fortunately the L series of lines shown in fig. 5.1, or even the M series for very heavy elements, have very suitable properties. Again it turns out that of the vast number of possible lines, L_{α_1} and L_{β_1} are far stronger than the remainder, which can normally be ignored. A similar effect narrows the M series of lines down to a few useful lines. Table 5.1 shows the energy and associated wavelength of the strongest K, L and M lines of the elements. The most efficient production of X-rays generally occurs when the bombarding electrons have about three times the X-ray energy. A study of table 5.1 will show that all elements have at least one strong X-ray line with energy less than 10 keV and therefore there should be no difficulty in analysing for all elements, using a scanning electron microscope capable of operating at 25–30 keV.

In addition to direct excitation by electrons there is a further mechanism of X-ray production which must be considered in micro-analysis. X-rays which are passing through a specimen (perhaps having been generated previously by electron excitation) can themselves excite atoms which then emit characteristic X-rays of a slightly lower energy. Thus, for example, in a brass specimen the zinc K_α X-rays (energy 8·64 keV) can excite extra copper K_α X-rays whose energy is less (8·05 keV, see table 5.1). This effect is known as *fluorescence*. It is not a very efficient process, since only a few per cent of the higher energy rays will successfully excite the lower energy radiation. However, it may significantly alter the relative amounts of characteristic radiation coming from alloys and compounds, particularly when elements with quite similar atomic numbers are present. In the example already quoted we would expect that if the composition of the brass were 70% Cu, 30% Zn we would find rather more than the expected proportion of X-rays emitted were Cu K_α and rather fewer than expected were Zn K_α, because of the fluorescence effect. We shall see later that fluorescence is one of the factors which makes accurate quantitative analysis very difficult.

One of the key factors which determines the scale on which micro-analysis can be carried out is the *interaction volume* or the volume of the specimen which is penetrated by electrons and from which X-rays may be emitted. As fig. 5.2 indicates, the primary electron beam is scattered throughout a pear-shaped region of a solid specimen before all the electrons have lost so much of their energy that they can no longer excite X-rays. It is difficult to make this volume much smaller than $1\,\mu m \times 1\,\mu m \times 1\,\mu m$ without reducing the energy of the electron beam so much that no useful X-rays are excited. As a consequence the smallest volume which it is practicable to analyse in a scanning electron microscope is $1\,(\mu m)^3$. Even though X-rays

Table 5.1.

Element	Atomic number Z	Relative atomic mass A_r	K_{α_1} E(keV)	K_{α_1} (nm)	L_{α_1} E(keV)	L_{α_1} (nm)	M_{α_1} E(keV)	M_{α_1} (nm)
			(*a*)	(*b*)	(*a*)	(*b*)	(*a*)	(*b*)
Hydrogen	1	1·0						
Helium	2	4·0						
Lithium	3	6·9	0·05					
Beryllium	4	9·0	0·11	11·40				
Boron	5	10·8	0·18	6·76				
Carbon	6	12·0	0·28	4·47				
Nitrogen	7	14·0	0·39	3·16				
Oxygen	8	16·0	0·52	2·36				
Fluorine	9	19·0	0·68	1·83				
Neon	10	20·2	0·85	1·46				
Sodium	11	23·0	1·04	1·19				
Magnesium	12	24·3	1·25	0·99				
Aluminium	13	27·0	1·49	0·83				
Silicon	14	28·1	1·74	0·71				
Phosphorus	15	31·0	2·01	0·61				
Sulphur	16	32·1	2·31	0·54				
Chlorine	17	35·5	2·62	0·47				
Argon	18	39·9	2·96	0·42				
Potassium	19	39·1	3·31	0·37				
Calcium	20	40·1	3·69	0·34	0·34	3·63		
Scandium	21	45·0	4·09	0·30	0·39	3·13		
Titanium	22	47·9	4·51	0·27	0·45	2·74		
Vanadium	23	50·9	4·95	0·25	0·51	2·42		
Chromium	24	52·0	5·41	0·23	0·57	2·16		
Manganese	25	54·9	5·90	0·21	0·64	1·94		
Iron	26	55·8	6·40	0·19	0·70	1·76		
Cobalt	27	58·9	6·93	0·18	0·77	1·60		
Nickel	28	58·7	7·48	0·17	0·85	1·46		
Copper	29	63·5	8·05	0·15	0·93	1·33		
Zinc	30	65·4	8·64	0·14	1·01	1·23		
Gallium	31	69·7	9·25	0·13	1·10	1·13		
Germanium	32	72·6	9·88	0·12	1·19	1·04		
Arsenic	33	74·9	10·54	0·12	1·28	0·97		
Selenium	34	79·0	11·22	0·11	1·38	0·90		
Bromine	35	79·9	11·92	0·10	1·48	0·84		
Krypton	36	83·8	12·65	0·10	1·59	0·78		
Rubidium	37	85·5	13·39	0·09	1·69	0·73		
Strontium	38	87·6	14·16	0·09	1·81	0·69		
Yttrium	39	88·9	14·96	0·08	1·92	0·64		
Zirconium	40	91·2	15·77	0·08	2·04	0·61		

Table 5.1.—*Continued.*

Element	Atomic number Z	Relative atomic mass A_r	K_{α_1}		L_{α_1}		M_{α_1}	
			E(keV)	(nm)	E(keV)	(nm)	E(keV)	(nm)
			(*a*)	(*b*)	(*a*)	(*b*)	(*a*)	(*b*)
Niobium	41	92·9	16·61	0·07	2·17	0·57		
Molybdenum	42	95·9	17·48	0·07	2·29	0·54		
Technetium	43	98·0	18·36	0·07	2·42	0·51		
Ruthenium	44	101·1	19·28	0·06	2·55	0·48		
Rhodium	45	102·9	20·21	0·06	2·70	0·46		
Palladium	46	106·4	21·17	0·06	2·70	0·44		
Silver	47	107·9	22·16	0·06	2·98	0·41		
Cadium	48	112·4	23·17	0·05	3·13	0·39		
Indium	49	114·8	24·21	0·05	3·29	0·38		
Tin	50	118·7	25·27	0·05	3·44	0·36		
Antimony	51	121·7	26·36	0·05	3·60	0·34		
Tellurium	52	127·6	27·47	0·04	3·77	0·33		
Iodine	53	126·9	28·61	0·04	3·94	0·31		
Xenon	54	131·3	29·77	0·04	4·11	0·30		
Caesium	55	132·9	30·97	0·04	4·29	0·29		
Barium	56	137·3	32·19	0·04	4·46	0·28		
Lanthanum	57	138·9	33·44	0·04	4·65	0·27	0·83	1·49
Hafnium	72	178·5	55·78	0·02	7·90	0·16	1·64	0·75
Tantalum	73	181·0	57·52	0·02	8·14	0·15	1·71	0·73
Tungsten	74	183·8	59·31	0·02	8·40	0·15	1·77	0·70
Rhenium	75	186·2	61·13	0·02	8·65	0·14	1·84	0·67
Osmium	76	190·2	62·99	0·02	8·91	0·14	1·91	0·65
Iridium	77	192·2	64·88	0·02	9·17	0·14	1·98	0·63
Platinum	78	195·1	66·82	0·02	9·44	0·13	2·05	0·60
Gold	79	197·0	68·79	0·02	9·71	0·13	2·12	0·58
Mercury	80	200·6	70·81	0·02	9·99	0·12	2·19	0·56
Thallium	81	204·4	72·86	0·02	10·27	0·12	2·27	0·55
Lead	82	207·2	74·96	0·02	10·55	0·12	2·34	0·53
Bismuth	83	209·0	77·10	0·02	10·84	0·11	2·42	0·51
Polonium	84	210·0	79·28	0·02	11·13	0·11	?	?
Astatine	85	210·0	81·50	0·02	11·43	0·11		
Radon	86	222·0	83·77	0·01	11·73	0·11		
Francium	87	223·0	86·09	0·01	12·03	0·10		
Radium	88	226·0	88·45	0·01	12·34	0·10		
Actinium	89	227·0	90·87	0·01	12·65	0·10		
Thorium	90	232·0	93·33	0·01	12·97	0·10	3·00	0·41
Protoactinium	91	231·0	95·85	0·01	13·29	0·09	3·08	0·40
Uranium	92	238·0	98·42	0·01	13·61	0·09	3·17	0·30

may be excited throughout this volume the number which reach the specimen surface and are emitted will depend both on the energy of the X-rays and the average atomic weight of the specimen. For example, soft (low energy, long wavelength) X-rays such as carbon K_α are readily absorbed by solids and therefore relatively few escape from the surface. On the other hand, hard (high energy, short wavelength) X-rays such as molybdenum K_α can penetrate many micrometres of most solids and are depleted only a very little by absorption in the specimen. Clearly, therefore, the volume which is being analysed depends very critically on

(*a*) the energy of the electron beam,
(*b*) the energy (wavelength) of the X-ray being studied
and (*c*) the average atomic weight of the specimen.

As was stressed earlier, this complexity makes accurate analysis extremely difficult.

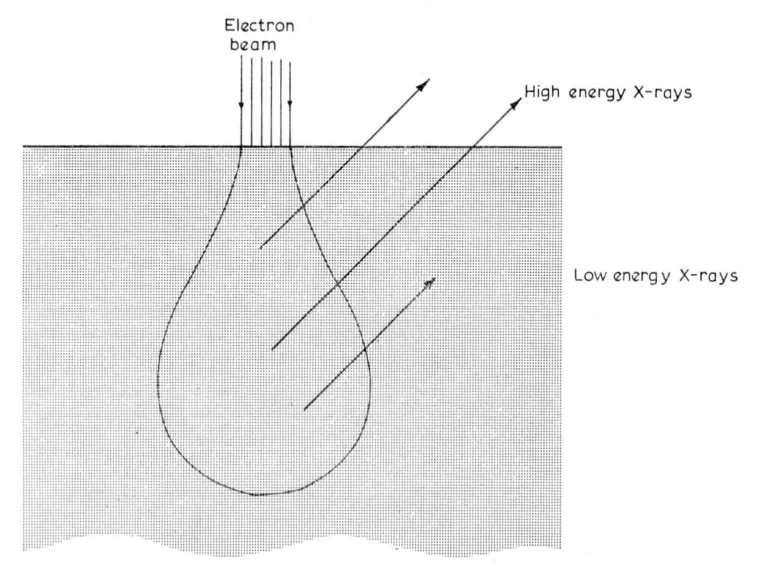

Fig. 5.2. The 'interaction volume' within a solid specimen from which X-rays are emitted. Low energy (soft) X-rays which are generated deep in the specimen may be absorbed before they reach the surface.

5.2. *Detection and counting of X-rays*

Ideally a system which is to be used for the analysis of a specimen in a scanning or transmission electron microscope should be able to reproduce for us the entire spectrum of X-rays emitted from the surface. Thus, if the specimen were a mixed oxide of chromium and iron, such as we might find on the surface of a stainless steel, we would expect its X-ray spectrum to look rather like fig. 5.3 (*a*). The characteristic ' lines ' should be very sharp, narrow peaks superimposed on a smooth background. The characteristic energy (or wavelength) of each peak enables us to identify it and we can therefore perform a qualitative analysis (i.e. determine what elements are present). Notice that there is only a single line characteristic of oxygen whereas the heavier elements Cr and Fe each show four distinct lines. From this complete spectrum we could choose the strongest line from each element—O K_α, Fe K_α and Cr K_α in this example—and compare its height (i.e. intensity) with that of a similar peak generated from a standard specimen of known composition. In this case we would probably use as standard specimens pure iron, pure chromium and a stable oxide such as alumina. The ratio of the heights of a peak from the specimen and from the standard then gives us a measure of the quantitative composition of the specimen (in this case the percentages of oxygen, iron and chromium present). If we had such an ideal system we could therefore analyse a specimen first qualitatively and then quantitatively for any of the elements present.

As we have by now learnt to expect, the ' ideal ' is rarely attainable in practice. In the present case we generally have to choose between two very different methods of obtaining some of the information shown in fig. 5.3 (*a*). Many electron beam instruments, particularly SEMs, are equipped with a *non-dispersive* (or *energy-dispersive*) detection system which is able to detect and display most of the X-ray spectrum shown in fig. 5.3 (*a*) but with some loss of precision, as indicated in fig. 5.3 (*b*). On the other hand, most purpose-built electron probe microanalysers incorporate one or two *wavelength-dispersive* spectrometers. These devices can determine extremely accurately the position of a single X-ray line (i.e. its wavelength or energy) and are particularly suited to measuring the height (intensity) of a peak above the background level. Well-equipped modern instruments have both types of detection system, so that the energy-dispersive system (EDS) can be used for qualitative analysis while the wavelength-dispersive spectrometer (WDS) is reserved for quantitative analysis of particular X-ray lines. We shall now consider the two types of analysis system in more detail, to see why there is still a need for both.

We shall deal with the energy-dispersive detection system first, since, although it is historically the more recent, it is now the more generally applicable and certainly the more versatile system. In outline the system normally consists of a small piece of semiconducting silicon which is held in such a position that as many as possible of the X-rays emitted from the specimen fall upon it. Since X-rays cannot be deflected and therefore will not follow a curved path the detector must be in the line of sight of the specimen. This means that in a

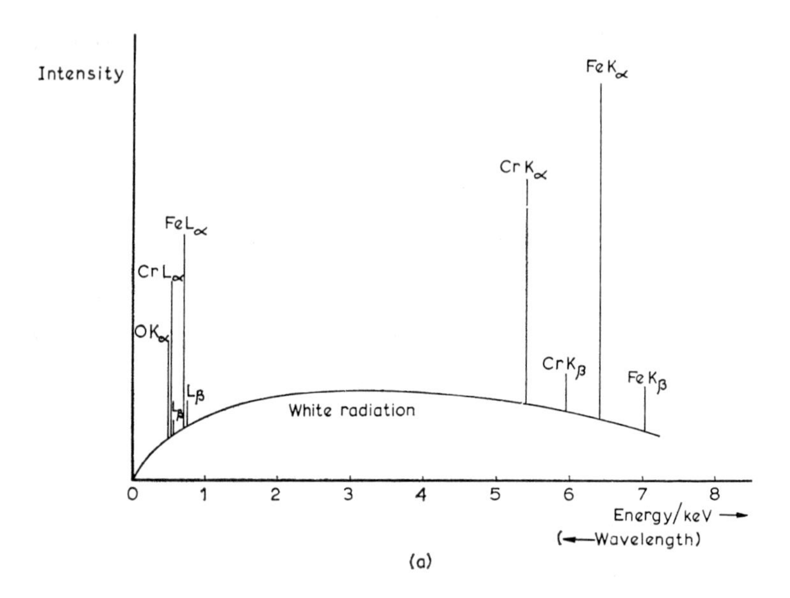

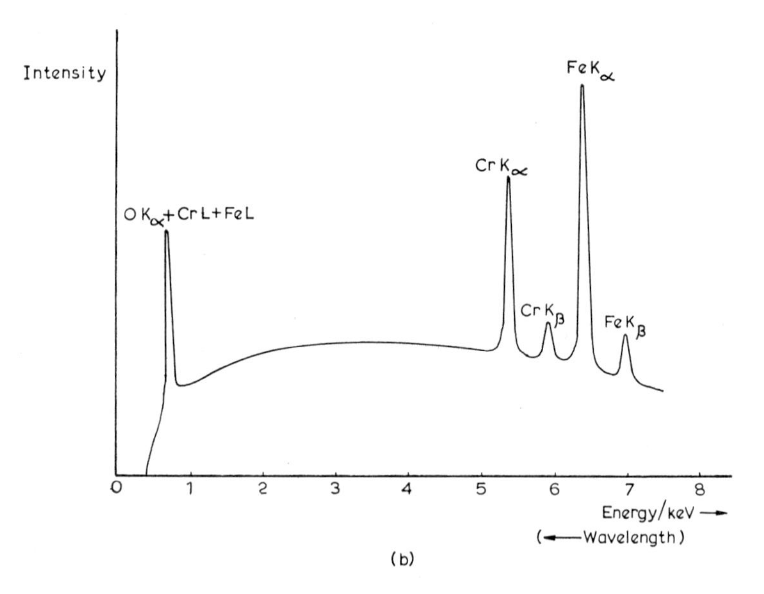

Fig. 5.3. (*a*) The X-ray spectrum which would be emitted from a mixture of iron and chromium oxides on electron bombardment. (*b*) The spectrum detected by an energy dispersive analysis system; the low energy lines cannot be resolved.

scanning electron microscope it normally occupies a similar position to the electron detector (cf. fig. 4.1). In order to collect as many X-rays as possible the silicon should be as near to the specimen as is practicable. In a SEM it may be possible to place the detector 20 mm or less from the specimen, but the problems are greater with a transmission electron microscope because of the essential presence of the objective lens very close to the specimen. However, in practice it has proved possible to install a semiconductor detector with an active area of 10–12 mm^2 at an acceptable distance from the specimen in virtually every model of SEM or TEM on the market.

The detector works in the following way. Each incoming X-ray excites a number of electrons into the conduction band of the silicon, leaving an identical number of positively charged holes in the outer electron shells. The energy required for each of these excitations is only 3·8 eV; consequently the number of electron–hole pairs generated is proportional to the energy of the X-ray being absorbed (detected). For example, an Al K_{α} X-ray, with an energy of 1·49 keV, will give rise to 392 electron–hole pairs. If a voltage is applied across the semiconductor a current will flow as each X-ray is absorbed in the detector and the magnitude of the current will be exactly proportional to the energy of the X-ray. In practice, if pure silicon is used the current generated is minute compared with the current which flows normally when a voltage is applied; in other words the resistivity is too low. This is overcome by the use of three strategems which combine to make the final detector seem rather more complicated than it really is. Figure 5.4 shows the result, and contains all the features which are common to virtually all energy-dispersive detectors. The natural conductivity of the silicon is lowered by (*a*) making the whole detector a semiconductor p–i–n junction which is reverse biased, (*b*) doping the silicon with a small concentration of lithium, often abbreviated to Si (Li), and (*c*) cooling the whole detector to liquid nitrogen temperature (−196°C). The current which normally passes between the gold electrodes is now very small indeed until an X-ray enters the detector, then the resultant current can be amplified and measured fairly easily. The detector shown schematically in fig. 5.4 consists of a Si (Li) semiconductor junction in which the *i* region occupies most of the 3 mm thickness. Thin layers of gold are necessary on both surfaces of the detector so that the bias potential can be applied. The film of gold on the outer face of the detector must be as thin as possible so that very few X-rays are absorbed in it; a layer only 20 nm thick provides adequate conductivity. The gold-coated outer surface must be further protected by a thin window, usually of beryllium. This window is necessary to prevent contaminants from the specimen chamber of the microscope from condensing on the very cold surface of the detector and forming a further barrier to the entry of X-rays. Unfortunately the window itself, despite being made of beryllium ($Z=4$) and only being a few micrometres thick, absorbs a significant proportion of the low energy X-rays falling on the detector and therefore makes light elements particularly difficult to detect.

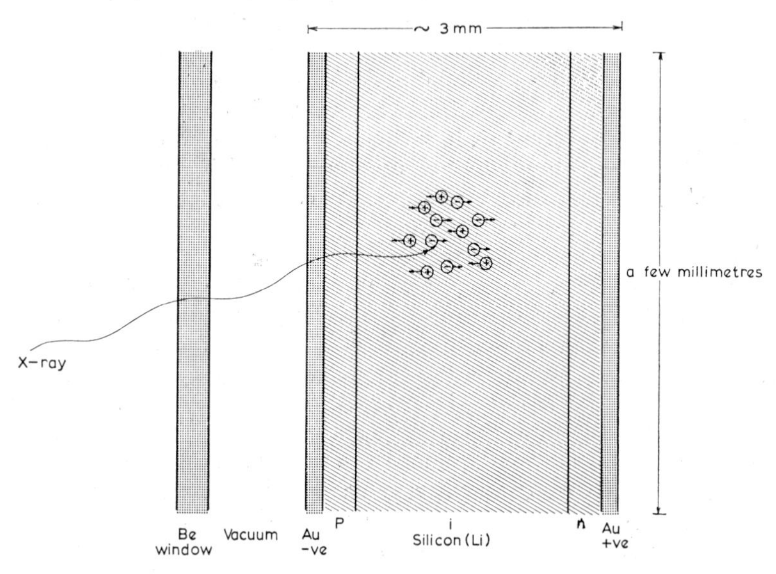

Fig. 5.4. A silicon energy-dispersive X-ray detector. The beryllium window and gold contact layers are grossly exaggerated in thickness; typical thicknesses would be 7–8 μm for the Be and 10–20 nm for the Au.

The current which flows between the electrodes when an X-ray enters the detector lasts for an extremely short time (less than 1 μs) and is normally referred to as a *pulse*. Each pulse is amplified and then passed to a multichannel analyser (MCA). This ' black box ' is capable of deciding which of perhaps 1000 channels, each representing a different X-ray energy, the pulse should be registered in. The MCA thus effectively collects a histogram of the energies of all the X-rays arriving at the detector; as indicated in fig. 5.5. This histogram is then displayed on a cathode-ray tube or television screen and although in reality composed of 1000 dots (one per channel) usually appears as a smooth curve similar to figure 5.3 (*b*). The MCA can *sort* incoming pulses very rapidly but obviously the whole system, detector, amplifier and MCA, takes a finite time to recover from one pulse before it can accurately process the next. This limits the rate at which X-rays can be counted; at the time of writing it is a simple matter to count at the rate of several thousands of counts per second (c.p.s.) and with more sophisticated electronics it is possible to deal with tens of thousands of c.p.s.

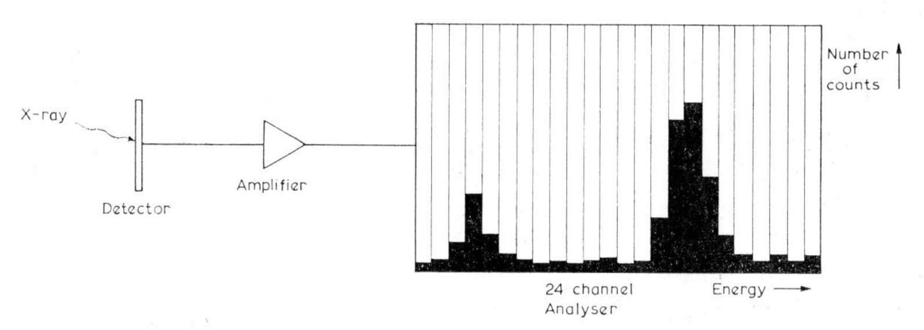

Fig. 5.5. A simplified energy-dispersive analysis system. Pulses from the detector are amplified and then stored in the appropriate channel of a multi-channel analyser. In reality the MCA would contain 500 or 1000 channels instead of the 24 shown here.

From the foregoing description it would seem that the EDS is an ideal system for presenting all the available X-ray data from a specimen in a convenient form from which both qualitative and quantitative analyses could be deduced. However, there are some serious limitations to be considered. One of the main disadvantages of a semiconductor detector is that the incoming X-ray must pass through a beryllium window, a gold electrode and a layer of ' dead ' silicon (the p-type region) before it can be detected. X-rays of very low energy (long wavelength) therefore have very little chance of being detected. In most cases it is impractical to use any X-rays of less than 1 keV for analysis. This eliminates the K_α lines of all elements lighter than sodium, including the frequently interesting elements carbon, oxygen and nitrogen. It may, however, be possible to improve this light element detection by arranging that the beryllium window is removed for short periods when the vacuum in the microscope is good enough. A further limitation of the detector is that it has a relatively poor ' energy resolution '; in other words, each X-ray line is not detected as a sharp line, as illustrated in fig. 5.3 (*a*), but as a broad peak, 100–200 eV wide, as indicated in fig. 5.3 (*b*). The consequences of the two effects (insensitivity to low energies and poor resolution) are evident in this figure. The Cr and Fe K_α and K_β lines are appreciably broader than in reality, while the Fe L_α and L_β lines cannot be separated. The lowest energy lines, including the important O K_α, cannot be seen at all. Clearly the information displayed in fig. 5.3 (*b*) does not indicate the composition of the specimen in the way that the true spectrum of X-ray emission would if only we could detect it and display it as shown in fig. 5.3 (*a*).

There are a few other disadvantages of EDS which must be borne in mind before we consider the undoubted advantages. The detector must be kept at 77 K at all times—clearly an experimental difficulty. Also, if quantitative analyses are required, the height of each peak above the background level is important. As is indicated in fig. 5.3 the background level tends to be enhanced by the EDS system and so the peak-to-background ratio which, as we shall see later, determines the limit of detectability of the analyser, is rather low.

Having painted such a gloomy picture of the energy-dispersive analysis system it is time to return to the positive side. One of the great advantages of a detector with no moving parts and a compact size is that it can be placed very near the specimen and it is therefore very efficient at collecting X-rays. The result of this is that the sort of spectrum shown in fig. 5.3 (*b*) can frequently be collected in only a few minutes. Consequently a qualitative analysis can be carried out quickly that will detect all but ten of the periodic table elements (and several of those ten are rarely needed—for example, neon, helium and lithium). As we shall see later, the quantitative accuracy of the analysis is often limited but even so a semi-quantitative estimate of the amount of each element present can often be made. The EDS therefore provides a comprehensive qualitative and semi-quantitative analysis in 5 to 10 min. This is 20 or 30 times faster than any other technique usable in an electron microscope.

The three areas in which the EDS system performs badly—light element detection, peak separation (i.e. energy resolution) and peak-to-background ratio—are the strong points of the other major X-ray detection system for electron microscopes, the wavelength-dispersive spectrometer (WDS). The principle of the WDS is that the X-radiation coming from the specimen is filtered so that only X-rays of a chosen wavelength (usually the characteristic wavelength of the element of interest) are allowed to fall on a detector. The ' filtering ' is achieved by a crystal spectrometer, which employs diffraction to separate the X-rays according to their wavelength. A common arrangement for the spectrometer is illustrated in fig. 5.6. The X-rays leaving the specimen at a certain angle ϕ, the *take-off angle*, are allowed to fall on to a crystal of lattice spacing d. If the angle between the incident X-rays and the crystal lattice planes is θ, then the only X-rays which will be diffracted by the crystal and thus reach the detector will be those obeying Bragg's Law (Section 2.7). The wavelength, λ, of the transmitted X-rays is therefore given by

$$\lambda = \frac{2d \sin \theta}{n}. \qquad (5.1)$$

If the spectrometer is to be used to detect, say, Fe K_α, then the characteristic wavelength (0·19 nm from table 5.1) is substituted into equation (5.1) and the appropriate value of θ can be calculated for the particular crystal in use. If the spectrometer is set to this angle then only the Fe K_α characteristic X-rays will reach the detector and be counted. The detector no longer has to discriminate between X-rays of different energies—it is sufficient for it to count the X-ray photons arriving. A much simpler detector than is found in the EDS can be used. This is the gas proportional counter, in which much faster counting rates can be tolerated. One of the disadvantages is that because the X-rays must be reasonably well collimated before reaching the crystal two sets of slits (S_1 and S_2) are generally used. In order that as many as possible of the X-rays leaving the crystal arrive at the detector the geometry of the spectrometer is chosen so that, as shown in fig. 5.6, all possible X-ray paths are focused on to the detector. In order to achieve this focusing effect the specimen, crystal and

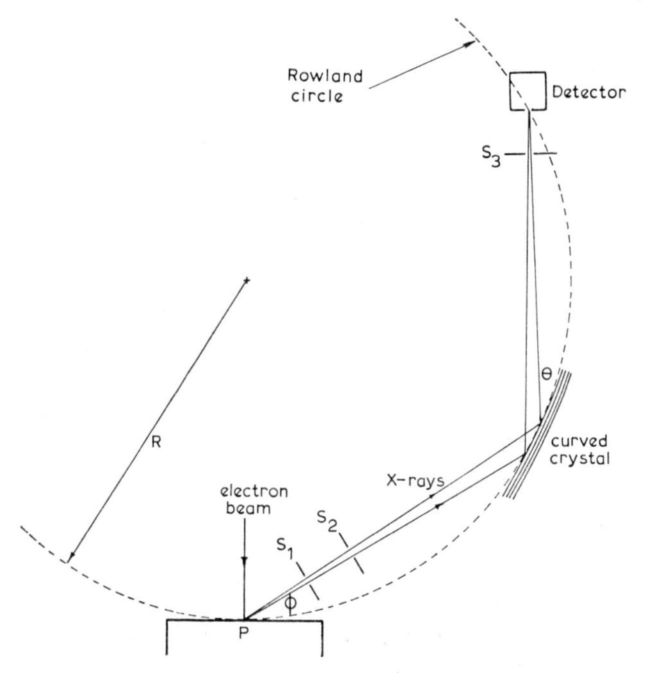

Fig. 5.6. A crystal X-ray spectrometer. X-rays emitted from the specimen are collimated by two slits S_1 and S_2, diffracted by the curved crystal and then focused on to the detector. For the maximum efficiency the specimen, crystal and detector must all lie on the Rowland circle of radius R.

detector must all be on a circle of radius R known as the Rowland circle and shown dotted in fig. 5.6. Also the crystal must be curved—in the Johan type of spectrometer shown in the figure it is bent to a radius $2R$. These requirements place several restrictions on the analysis system: the spectrometer is necessarily quite large; its mechanism is complicated, since it has to be able to alter θ while keeping both the crystal and the detector on the Rowland circle; precision engineering is required, since, in order to discriminate between very close lines, the angle θ must be set to an accuracy of better than one minute of arc; finally the position of the specimen is absolutely critical since if it lies off the Rowland circle by as little as a few micrometres the number of X-rays reaching the detector will be severely reduced. These restrictions make the design of an X-ray spectrometer difficult and expensive, particularly if other convenient features are to be incorporated, such as facilities for changing the crystal and the detector without breaking the vacuum system of the microscope together with a design which allows the take-off angle, ϕ, to remain constant whatever the value of θ.

Despite the difficulty and expense of a crystal spectrometer it remains an essential tool for analysis in the electron microscope because of four major advantages. (1) The resolving power of a spectrometer with an appropriate crystal is very good. This implies that the X-ray lines appear almost as sharp as shown in fig. 5.3 (*a*) and there is rarely a problem of adjacent lines overlapping. (2) The peak-to-background ratio of each line is much higher (often ten times higher) than can be achieved with a solid state detector. (3) Since the X-ray detector normally used in a spectrometer is capable of counting at very high rates (perhaps 50 000 counts per second) it may be possible to collect data from a single element in a very short counting time. (4) The most compelling reason for using a spectrometer, however, is its ability to detect X-rays from light elements. With suitable crystals of very large lattice spacing it is possible to detect and count X-rays as soft as boron K_α or even beryllium K_α. The spectrometer is therefore able to fill in the ‘ gap ’ at the low energy end of the X-ray spectrum which the EDS cannot detect. Using a spectrometer, therefore, analysis for carbon, nitrogen and oxygen can be performed routinely.

On the other hand, the disadvantages of an X-ray spectrometer make its use very time-consuming and hence restrict its major application to the things it does particularly well—detection of light elements and quantitative measurement of the heights of single peaks. One factor which tends to make qualitative analysis difficult is the proliferation of X-ray lines seen by the spectrometer. Since it employs diffraction there will be not one angle of the crystal (θ) at which a certain line is detected but several, one corresponding to each value of n in equation (5.1). Thus there may be as many as seven or eight *orders of reflection* detectable from a major X-ray line and therefore the spectrum contains far more lines than that collected by a semiconductor detector. As well as making identification of any individual line more difficult this effect also means that the angle of the spectrometer, θ, must be set very carefully for each analysis and re-set again equally carefully for a ' background ' measurement. It is this continual delicate ' tuning ' of the instrument which contributes to the long time necessary for the analysis of each element.

In summary, the two main X-ray detection and measurement systems, the semiconductor detector (EDS) and the crystal spectrometer (WDS), each have their strong points. The EDS is clearly better suited to rapid qualitative analyses while the WDS may give a more accurate quantitative result for any particular element and is essential for the analysis of elements lighter than sodium. As we shall see in the next section, both have wide application in electron microscopy and a really versatile analysing microscope really needs both.

5.3. *Analysers based on the scanning electron microscope*

The earliest *electron probe microanalysers* were simply machines which generated a beam of electrons, directed it at a specimen and used one or two crystal spectrometers to analyse the X-radiation emitted. This represented a great advance in analytical technique since the electron beam could be directed at a relatively small part of a solid specimen and an analysis obtained without destroying the specimen. The region of the specimen to be analysed was selected using a light microscope and the accuracy with which the beam could be positioned was rather limited. It was soon realized that by scanning the electron beam and installing an electron detector an imaging system could be incorporated. At this time SEMs were not highly developed and the electron imaging system was usually a secondary consideration, the primary function of the machine being to detect X-rays as efficiently as possible. Since the size of the interaction volume discussed in Section 5.1 implies that no useful analysis can be made of a region smaller than $1\,\mu m \times 1\,\mu m \times 1\,\mu m$ there seemed to be no point in making the electron beam diameter any smaller than about $0{\cdot}5\,\mu m$ and hence the image resolution could never be better than this. The ' conventional ' microprobe analyser is therefore an instrument capable of directing at a solid sample a beam of electrons with a diameter of a little less than a micrometre. The current carried by the beam needs to be quite large, perhaps as much as $10\,\mu A$, since the crystal spectrometers used on this sort of machine are relatively inefficient (cf. Section 5.2) and a large number of X-ray photons must be generated in the specimen. The instrument can generally present an electron image of the region being analysed, although this is at a rather poor resolution compared with that which is now taken for granted on most scanning electron microscopes. An instrument of this type is shown in fig. 5.7: the large enclosures for the spectrometers are visible at either side of the electron-optical column. In this particular machine the electron beam is horizontal and the specimen is held vertically in a holder at the front of the console.

The great advantages of this type of 'conventional' analyser are the accuracy of its spectrometers and their good peak-to-background ratio (the benefits of this will appear in Section 5.5 when quantitative analysis is discussed), the wide range of electron beam currents available, its capability for analysing light elements, and the ease and efficiency of operation of a machine designed solely as an analysing instrument.

An alternative approach to obtaining at least a qualitative analysis, which has been much in vogue in the early seventies, is to mount an EDS system on to an existing scanning electron microscope. For reasons which are discussed in Section 5.5, this is not an ideal combination for the purposes of quantitative analysis but it is absolutely invaluable in extending the range of qualitative information available to the microscopist. The scientist looking at his specimen in the SEM can suddenly say "Stop! That is an unusual feature; what is it?" and very often a qualitative answer is all that is needed.

The semiconductor X-ray detector is very easily fitted to most SEMs, since they tend to have a large specimen chamber to accommodate the electron detector, large specimens, manipulation devices, etc. The usefulness of the EDS technique is now so widely recognized that almost half the SEMs purchased are ordered complete with a semiconductor detector. A typical instrument is illustrated in fig. 5.8. The success of this type of combination instrument has led to the development of a new breed of truly versatile analysers which achieve the high analytical standards of a 'conventional' microprobe analyser while retaining the high resolution imaging of a sophisticated scanning electron microscope. An example of this class of instrument is shown in fig. 5.9. This machine, for instance, is capable of presenting an SEM image at a resolution of 10 nm, is fitted with crystal spectrometers which can cover the element range from boron ($Z=5$) to uranium ($Z=92$), and has an EDS semiconductor detector permanently installed. With this sort of instrument any analytical problem on a fine scale can be followed up from the initial 'look' at the surface, through the qualitative analysis to find 'what is there', to the final quantitative analysis for elements of particular interest.

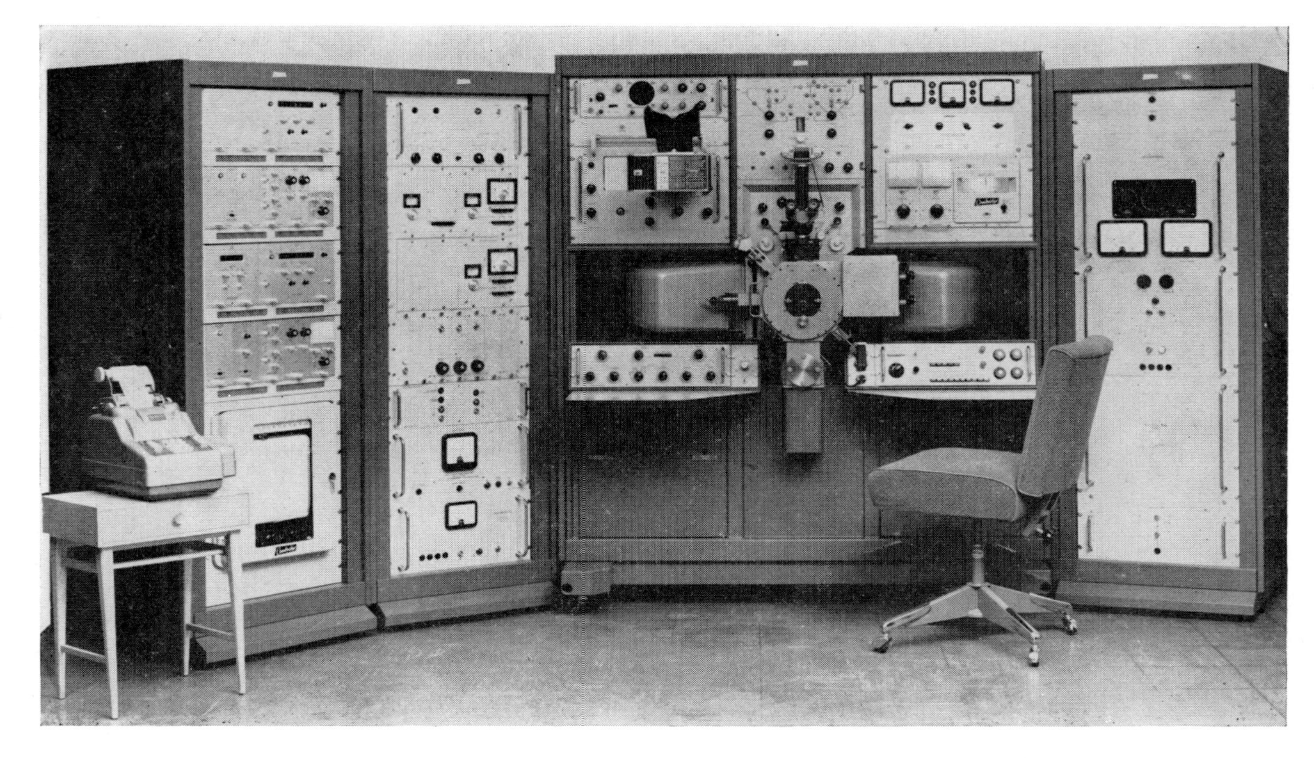

Fig. 5.7. An electron probe microanalyser, the Geoscan (Cambridge Scientific Instrument Co.).

Fig. 5.8. A scanning electron microscope fitted with an energy-dispersive X-ray detector.

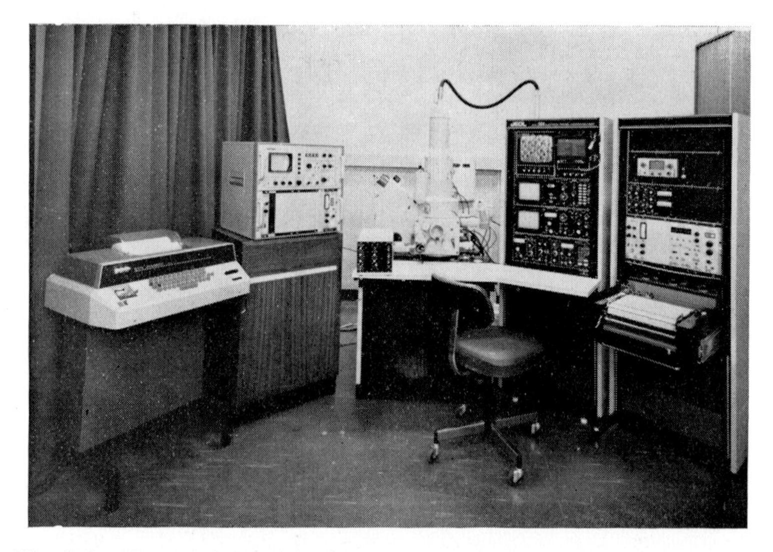

Fig. 5.9. A modern combined scanning microscope and probe analyser, the JXA-50A (JEOL Ltd.).

For all the instruments discussed in this section, which are based on a scanning electron beam, there is a very interesting variety of ways in which qualitative analytical information can be presented. We will now consider some of the more useful techniques. The simplest, most obvious and widely used of these is the ' spot ' analysis, in which the electron beam is stopped and positioned carefully on the point to be analysed, which has been selected on the SEM screen while the image was still being displayed. The X-ray data, either a single line being detected by a spectrometer or the whole spectrum being accumulated by an EDS, can be collected for as long as is necessary and the composition at the selected point (or strictly in a volume of about $1\,(\mu m)^3$) can be determined. This is extremely useful but can sometimes be a little *too* specific. If there are composition variations across the specimen we need to analyse a great number of such points to get a clear idea of the overall composition of the specimen, say on a $100\,\mu m \times 100\,\mu m$ scale. If such an overall analysis is required, perhaps as the first stage in the study of a specimen, it is far more appropriate to allow the electron beam to continue to scan its raster on the specimen while the X-rays are being collected. An average analysis of the whole region being scanned (i.e. the whole picture seen on the screen) is then obtained. This technique is an ideal one for the EDS system but cannot be carried out effectively with a WDS. The reason for this is that only one point (strictly, one line) of the specimen lies exactly on the Rowland circle (fig. 5.6). Consequently the remainder of the scanned area, although it emits X-rays, does not contribute so effectively to the analysis since those X-rays emitted from regions to the right and left of P in fig. 5.6 will not be focused on to the detector (i.e. defocusing occurs). For the analysis of the mean composition of a region of the specimen bigger than about $5\,\mu m$ square the EDS is therefore the only practical technique.

Once a qualitative analysis has been carried out, either from a point or an area of the specimen, it is possible to consider the distribution of any single element within the sample. Apart from the laborious, but most accurate, method of analysing a large number of separate points by either EDS or WDS, there are two techniques in which the special advantages of the scanning microscope can be used to advantage. If an instantaneous measure of the concentration of an element at a particular point on the specimen can be found, then it is possible to display this concentration as a function of position on one c.r.t. screen *while* the electron beam is scanned across a line of the specimen. It is quite simple to install a *ratemeter* in the X-ray counting system in order to achieve an ' instantaneous measure '; in the WDS this needs to indicate the rate of counting of the detector since only X-rays from a single element will be arriving, but in the

EDS it is necessary to make the ratemeter sensitive only to the counts going to a selected number of channels. The two arrangements are shown schematically in fig. 5.10. If the X-ray count rate measured by the ratemeter is used to deflect a spot on the c.r.t. screen then the sort of trace of composition versus distance illustrated in fig. 5.11 can be obtained. Notice that although for ease of recognition the ' line ' of the analysis is shown superimposed on a micrograph of the specimen, these must be taken separately. First an ordinary electron image of the specimen is recorded, then the c.r.t. spot is scanned along the chosen line and this line image is superimposed on the same photographic exposure, and finally the spot is made to travel very slowly along this line (perhaps taking 500 or 100 s) while the trace shown in fig. 5.1 (*b*) is displayed and photographed. As can be seen from the example illustrated, this technique makes it quite simple to relate the ' high spots ' where a particular element is concentrated on the features visible in the micrograph. The technique is difficult to make really quantitative since the electron beam does not normally spend long enough on each spot of the specimen to excite enough X-rays for an accurate analysis. The c.r.t. trace is therefore usually presented, as here, with no vertical scale indicated and is only assumed to give a rough guide to the variation in composition across the specimen. However, this is often extremely useful.

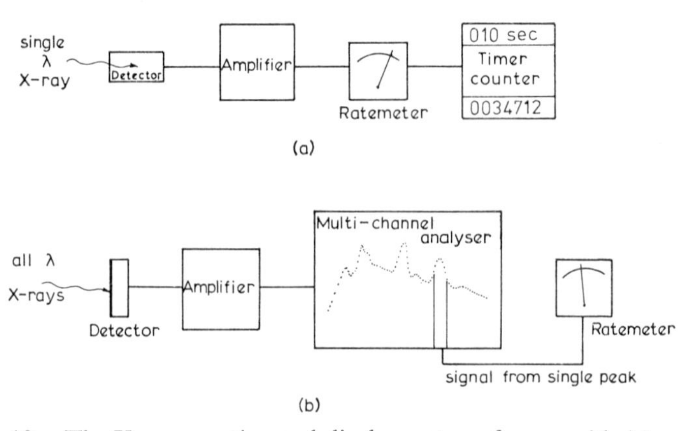

Fig. 5.10. The X-ray counting and display systems for use with (*a*) a crystal spectrometer and (*b*) an energy-dispersive detector.

A variant of this ' line ' technique which is applicable to longer lines can be achieved if the electron beam is held stationary while the specimen is driven (by a small motor) at a constant speed beneath it. The ratemeter output can then be displayed as the trace of a pen on a

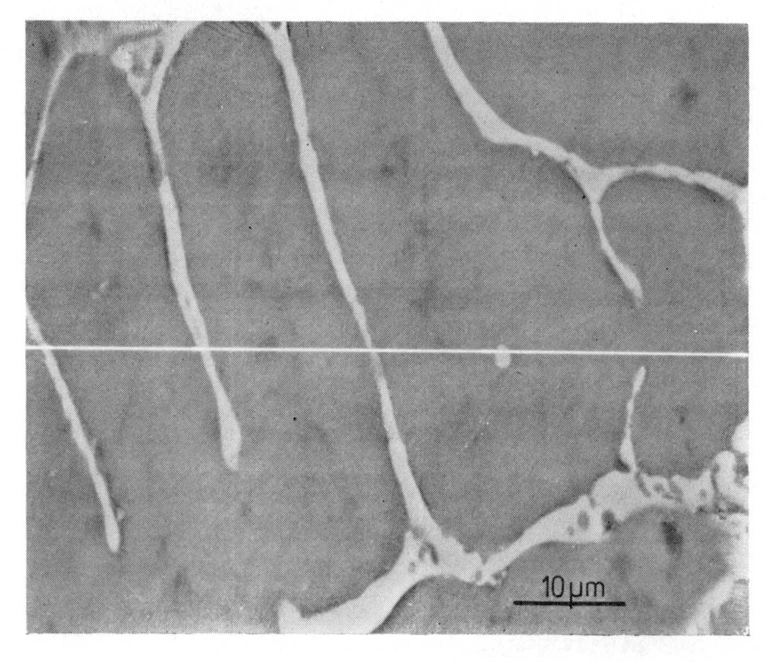

(*a*)

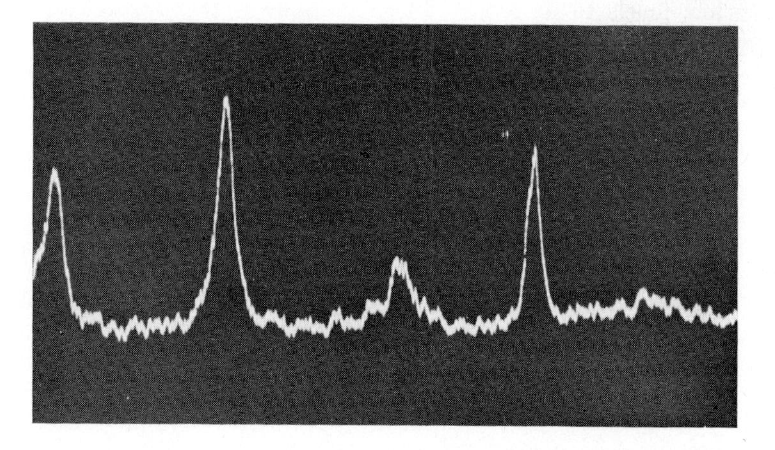

(*b*)

Fig. 5.11. The variation of copper concentration across a section of an aluminium–copper alloy. The trace in the lower part of the figure is the copper count rate (i.e. the output of the ratemeter) as the electron beam was scanned along the line indicated on the micrograph.

chart recorder, which can be as long as is needed. This type of analysis is particularly useful when a WDS is being used, since, if the electron beam were scanned along a line any longer than about 5 μm, the 'defocusing' effects described earlier would occur and a false picture of the element distribution would be obtained.

There is a second way of displaying information about the distribution of a single element within a sample. If the X-ray counter (of either a WDS or an EDS system) is used in a similar way to the electron detector of the conventional SEM then a concentration map of the specimen surface can be produced on the c.r.t. screen. The display is not very subtly graded: every time an X-ray is counted the spot on the display c.r.t. is made bright; as it is scanned across its usual raster bright dots therefore appear at positions corresponding to the presence of the selected element. To a first approximation the closer these dots are the greater must be the concentration of element in that region of the specimen and the brighter the image will appear. An example of this type of 'X-ray mapping' display is shown in fig. 5.12. The usefulness of this sort of display is obvious but, as might be expected, there are some snags. At low instrument magnifications, where large areas are being scanned, the WDS cannot be used because of the defocusing effect and here again the EDS has an advantage. However, neither detection system is particularly effective at giving enough X-ray counts above the background level to enable the distribution of elements present at very low levels to be seen. The technique is therefore effectively limited to cases where the local concentration of the element of interest is at least 10%. Even at these levels the electron beam must be scanned rather slowly across the specimen and it is not uncommon to take photographs with a 1000 s exposure (almost 17 min!). Nonetheless the X-ray mapping technique is a very graphic way of presenting qualitatively the distribution of one or several elements on a scale of tens of micrometres or above.

When using any of the methods of displaying analytical results which have just been discussed it is necessary to bear in mind some practical problems which may be encountered. Since the specimen is to all intents and purposes in a scanning electron microscope it must have a conducting surface. Non-conducting specimens must therefore be coated before being examined. The coating, however, will absorb X-rays as they are emitted from the specimen and will also emit its own characteristic X-rays. Consequently the coating should be as thin as possible, of as low an atomic weight as possible, and it must not contain an element which might be of interest in the specimen. For these reasons it is not particularly sensible to use the gold or gold–palladium alloys generally chosen for SEM work, since they consist of very heavy atoms (see table 5.1). It is better practice to coat the specimen with carbon or aluminium, even though the electron image may not look quite as good, since more X-rays will then escape to be analysed.

Another point is sometimes forgotten when the probe micro-analyser is based on an SEM; although the electron image of the specimen has such a great depth of field that rough specimens can be examined, these same specimens may give rise to misleading analytical results. The major reason for this is that although the secondary electrons which create the image can travel in curves (e.g. fig. 4.7) the X-rays emitted from the same points can only travel in straight lines. As fig. 5.13 shows, this may mean that although we can ' see ' a certain part of the specimen we are unlikely to detect X-rays from it and we may therefore be tempted to say that it is a region of low concentration of the element we are considering. As an extreme example of this effect consider the copper oxide spheres shown in fig. 5.14; the electron image (fig. 5.14 (*a*)) makes it clear that they are homogeneous spheres whereas the oxygen X-ray image (fig. 5.14 (*b*)) would imply that there is a variation of oxygen content across each sphere. In fact this appearance arises because the left-hand side of each sphere faces the X-ray detector whereas the other side is hidden from it.

Another misleading situation can arise if we forget that the X-ray analysis arises from the whole of the ' interaction volume ' discussed in Section 5.1 (fig. 5.2). It is quite possible that X-rays could be arriving at the detector from regions of the specimen which we cannot ' see ' or which we think are not being hit by the electron beam. Figure 5.15 illustrates three of the more common ways in which this can happen. It is fairly obvious that the situations illustrated in (*a*) and (*b*) will arise, and X-rays from sub-surface phases or from across boundaries may be detected. This effect has already been seen in fig. 5.11, where the apparent thickness of the copper-rich region as displayed on the X-ray trace is far greater than the real thickness seen in the secondary electron image. The third possibility (fig. 5.15 (*c*)) needs further explanation. Two important types of high energy radiation are emitted from the specimen while it is being analysed: ' reflected ' electrons and characteristic X-rays. Since they are both emitted in all directions it is possible—even likely—that X-rays will be detected from other parts of the specimen, as shown in the diagram, or from the specimen chamber of the microscope itself. Whether they are excited by reflected electrons or are fluoresced by X-rays from the specimen, these ' secondary ' X-rays will clearly be misleading if they reach the detector, since they do not come from the region of the specimen which is supposedly being analysed and may indeed indicate the presence of quite incorrect elements. The WDS is less prone to this sort of error than the EDS since the X-rays reaching the detector

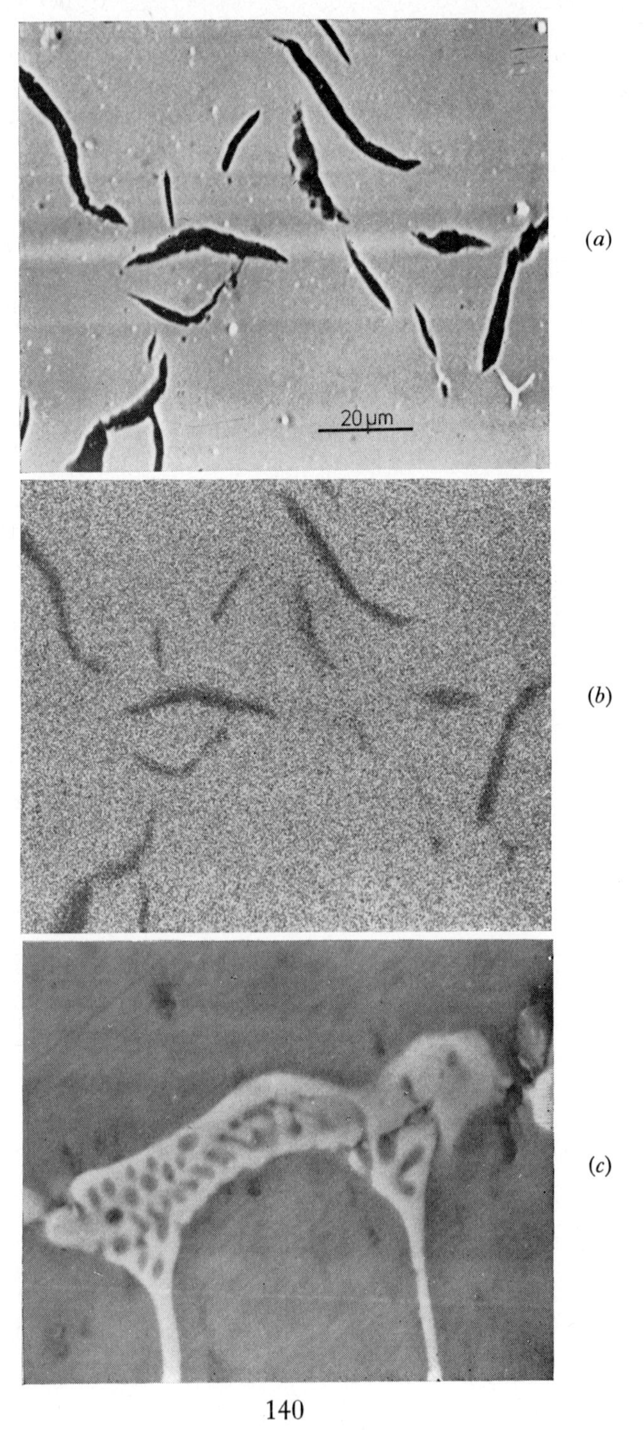
20 μm
(a)
(b)
(c)

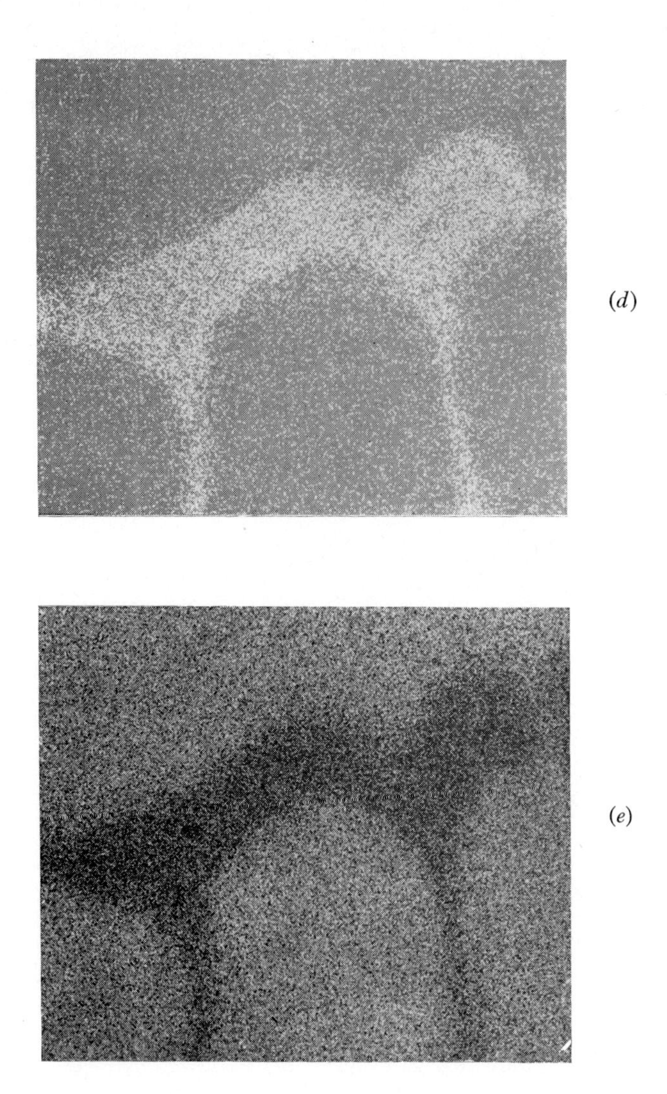

Fig. 5.12. X-ray distribution maps from two alloys. (*a*) Electron image of a cast iron showing the graphite flakes; (*b*) iron K_α distribution map; (*c*) a small region of the aluminium–copper alloy shown in fig. 5.11, electron image; (*d*) copper K_α map; (*e*) aluminium K_α map. Despite the obvious presence of fine details within the lighter phase in (*c*) the X-ray distribution maps do not show it because their resolution is limited to 1 μm.

have been quite severely collimated by the slits shown in fig. 5.6 and very few secondary X-rays are able to get to the detector. However, most semiconductor detectors have no such collimation and therefore are liable to give a misleading result, particularly from very rough specimens.

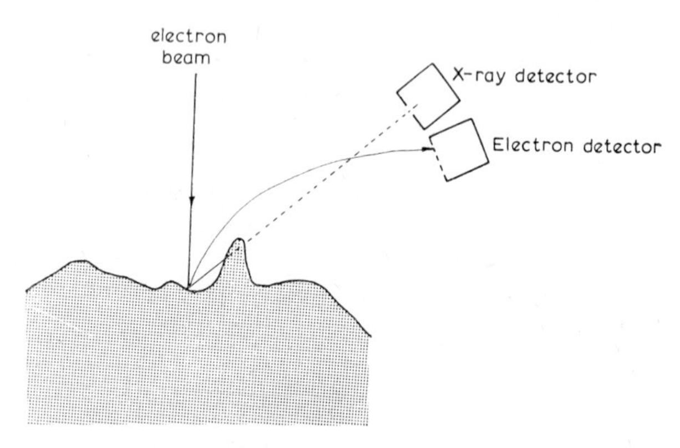

Fig. 5.13. X-ray imaging from a rough specimen. The point indicated here will be seen in an electron image because the secondary electrons can travel in a curved path to the detector. The same point will appear to have no X-rays emanating from it, however, since they can only travel in a straight line to their detector.

5.4. *Analysers based on the transmission electron microscope*

It would obviously be very useful if unknown features within transmission electron microscope specimens could be analysed in one of the ways just described for solid specimens. However, there are three difficulties which make it impracticable to do this simply by incorporating an EDS or WDS detector near the specimen. Briefly these are that very few X-rays will be generated in a thin specimen, there is very little room for them to be extracted from the microscope column, and the electron beam in a TEM is of such a large diameter that usefully small features could not be analysed. It was not until the early 1970s that all three difficulties could be overcome simultaneously. Microanalysis in a TEM is now becoming more widespread because it is the only way in which some types of analytical problems can be tackled.

Very little can be done about the small number of X-rays generated in a thin specimen—there is only a certain small volume of material and an electron beam of a finite brightness (Section 2.3) can therefore only generate a limited number of X-rays. On the other hand, the thinness of the specimen means that smaller features can be analysed

(*a*)

(*b*)

Fig. 5.14. Oxygen K_{α} distribution map from oxidized copper spheres. The impression given by the map is that one side of each sphere has a higher oxygen content than the other. This arises simply because the X-ray detector (a crystal spectrometer in this case) is positioned so that it can ' see ' one side of each sphere whereas the other is in shadow.

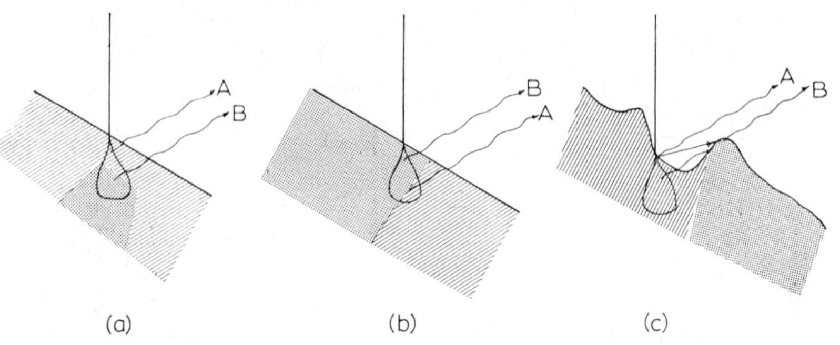

Fig. 5.15. Three common situations in which a false impression of local composition can be given. (*a*) In a region apparently consisting entirely of A, a region of B below the surface gives off B X-rays. (*b*) Near the boundary between phases A and B both types of X-ray are excited although the electron beam appears to be on the B phase. (*c*) Even farther away from a phase boundary in a rough specimen B X-rays can be excited by fluorescence from the A region.

than in an SEM-based analyser. The reason for this is illustrated in fig. 5.16. Not only is the region from which the X-rays are emitted (the interaction volume, Section 5.1) smaller in the direction of the electron beam but it is also narrower because there is less spreading of the beam within the specimen. This represents one of the main advantages of the TEM-based analyser—smaller regions can be analysed. The only way to make up for the small number of X-rays given off from the specimen in the TEM is to make the detector more *efficient*, in other words to make sure that more of the X-rays enter

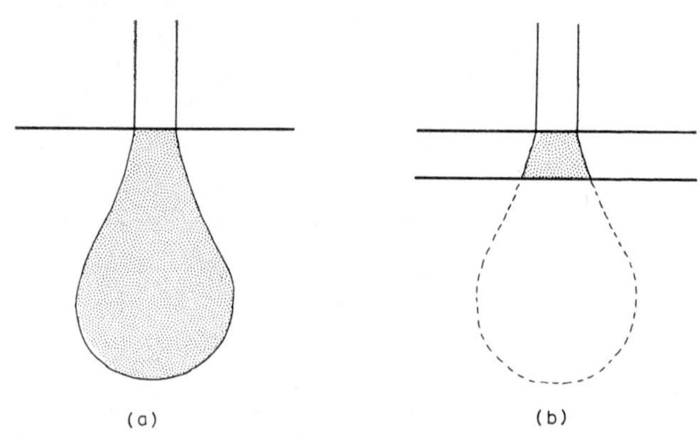

Fig. 5.16. A comparison of the size of the region from which X-rays are excited in (*a*) a solid specimen, and (*b*) a thin specimen.

the detector and are counted. Just about the only way to do this is to bring the detector nearer to the specimen and this brings us to the second problem; the specimen is generally positioned inside the casing of the objective lens, which may be 20 cm across (see, for example, fig. 5.17). There is very little space between the final condenser lens and the objective lens and therefore it is not easy to make a passage for the X-rays to travel to the outside of the microscope column to reach the detector.

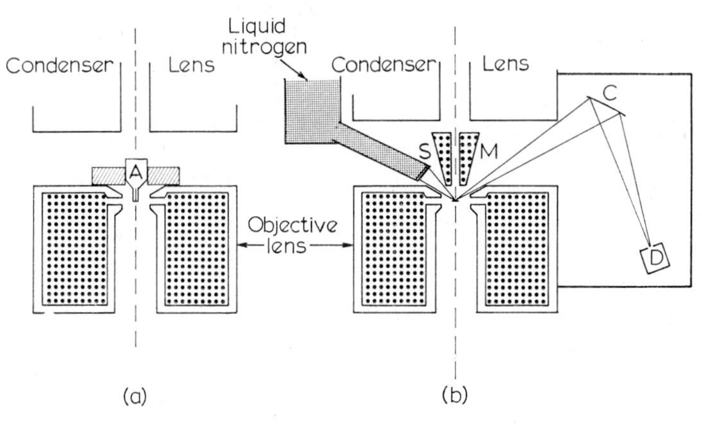

Fig. 5.17. X-ray detectors attached to a transmission electron microscope. (*a*) The specimen in a 'top-entry' specimen holder in a conventional microscope. (*b*) The modifications necessary for analytical microscopy; the specimen is mounted in the much smaller 'side-entry' holder, a mini-lens (M) is installed and there is now room for a silicon detector (S) and one or two crystal spectrometers (C).

In some microscopes it is possible to make the best of a bad job and position a detector so that a reasonable number of X-rays fall on it, without modifying the microscope column too much. However, a more radical approach has been used in the only commercial instrument which was purpose-built to be an electron microscope microanalyser (EMMA). This involves using an extra lens of a very different shape to the normal electron microscope lens, called a mini-lens. This, as its name implies, is much smaller than a conventional lens, and results in a very much greater space being available around and above the specimen, as fig. 5.17 illustrates. The mini-lens acts as a final condenser lens and the increased working space is such that even a crystal spectrometer can be brought close enough to the specimen to be used efficiently. At the same time the bulky top-entry specimen holder shown in fig. 5.17 (*a*) (and in fig. 3.2) must be replaced by a more open side-entry type which is basically a long rod which would run perpendicular to the paper in fig. 5.17 (*b*). One example is shown in fig. 5.18.

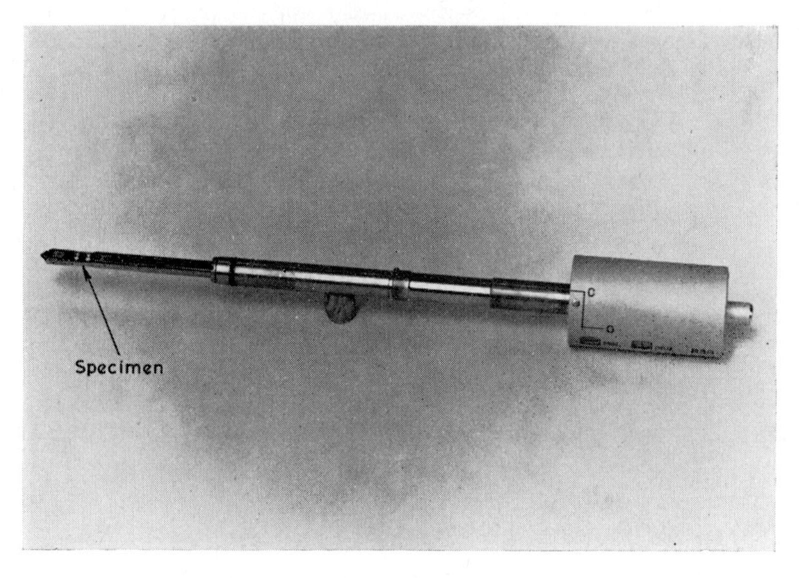

Fig. 5.18. A photograph of a side-entry specimen holder for a transmission electron microscope. Compare with the top-entry holders shown in fig. 3.2.

The mini-lens helps to overcome the third difficulty mentioned above. The minimum beam diameter produced by the normal condenser lenses of a microscope is between 0·5 and 5 μm. The addition of the mini-lens can reduce this to 0·2 μm or less and therefore opens up the possibility of analysing sub-micrometre features. Another effective way of achieving a small beam diameter at the specimen is to use a very strong objective lens such that the magnetic field above the specimen (at A in fig. 5.17 (*a*)) acts as a final condenser lens and reduces the spot size. This has the advantage that a very small beam diameter can be produced, but does not create any additional space above the specimen for the installation of an X-ray spectrometer. There are, therefore, two major methods of analysis using a TEM; either a semiconductor detector (EDS) is fitted to a fairly standard TEM, preferably with a mini-lens or an over-excited objective lens, or a crystal spectrometer is built into a special microscope with a mini-lens. The former approach is being adopted widely and is extremely useful for qualitative analyses; it is also convenient for use in those TEM instruments which can be used in the scanning transmission mode (STEM, see Section 4.4). The latter approach, involving the building of fully focusing spectrometers on to a special microscope, has led to the development of EMMA, illustrated in fig. 5.19. This is an instrument more suited to quantitative analysis, although as will be discussed in the next section, its accuracy is likely

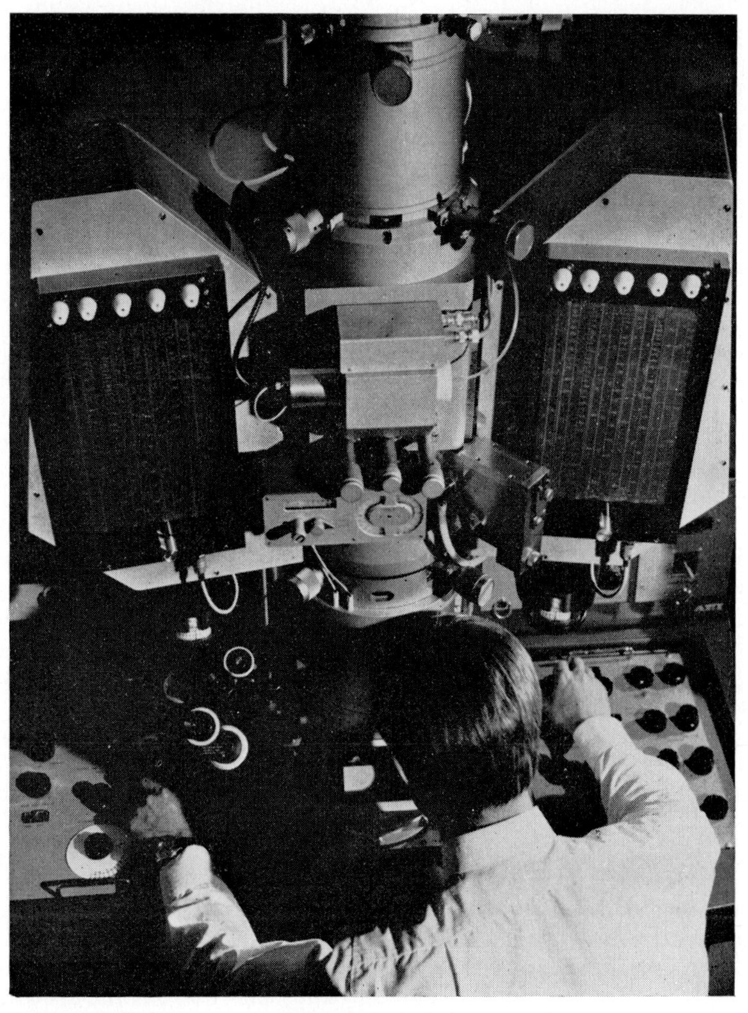

Fig. 5.19. A photograph of the analytical electron microscope EMMA-4 (AEI Scientific Apparatus Ltd.).

to be limited by the small number of X-rays emitted and by the problem of finding suitable standard specimens.

In any transmission microscope which does not form its image by scanning the beam, and of course this includes all conventional TEM instruments, we cannot hope to present analytical information in the great variety of ways described in Section 5.3 for SEM-based analysers. The sole technique for analysis, therefore, is to study the area of interest at an appropriate magnification using a ' large ' beam diameter, select the region of interest and condense the beam on to it using the mini-lens or the normal condenser lens system. The analysis of the selected spot is then made by collecting the X-ray spectrum over a measured period of time.

In a STEM instrument, of course, the full range of line scans and ' mapping ' displays of analytical information described earlier is available—yet another advantage of STEM over the conventional TEM.

There are one or two problems associated with TEM-based analysis which have to be weighed against the great advantages of a small selected area of analysis. The worst problem is usually the low X-ray count rate and, consequently, the poor detectability for small concentrations of an element. Another major disadvantage is that specimens must be prepared in such a way that their chemical composition is unaltered—this is a big problem for the biologist. A further troublesome effect is the fluorescence or backscattered electron excitations of X-ray from the specimen holder and microscope column. The specimen in a TEM is much more closely surrounded by metal objects than is usually the case in an SEM and mechanisms such as that illustrated in fig. 5.15 (*a*) therefore occur frequently. These problems can be alleviated, but not entirely cured, by using a specimen holder, and specimen support grids, made of carbon, beryllium or a polymer such as nylon. They then only emit very soft X-rays which cannot normally be detected and therefore do not interfere with the analysis.

Applications of TEM-based analyses in biological science illustrating the use of both EDS and WDS techniques are described in Section 5.6 at the end of this chapter.

5.5. *Quantitative analysis in an electron microscope*

In all the discussion of analysis up to this point we have been concerned only with a qualitative answer to the question " what is present in that region of the specimen? " There are many occasions when it would be highly desirable to know more about the specimen in a quantitative way. For instance, three questions which are often of importance are: " What is the size and shape of the region which has been analysed? ", " What is the smallest amount of element X which could be detected? " and " How much of this element is present in this part of the specimen? " To find even an approximate answer to any of these questions requires a far deeper consideration of what happens inside the specimen as the X-rays are first generated, and then partially absorbed again, than we have so far had space for. All we can hope to achieve in this section is some understanding of why these questions are difficult to answer precisely and what sort of order of magnitude the answers are likely to have.

Let us take the *volume analysed* first: all we can say is that it is that region from which the X-rays which reach the detector are emitted. As we have seen at the beginning of this chapter this must depend on where the X-rays are generated (i.e. how far the electron beam penetrates the sample) and how strongly the X-rays are absorbed by the specimen on their way out. The volume is therefore going to depend on the electron beam energy, the average atomic weight of the sample, the wavelength of the characteristic X-rays being studied, the absorption coefficients of the specimen for these X-rays and the angle of incidence of the electrons on the specimen surface, to mention only the five most obvious factors. Another confusing point is that, of the X-rays emitted, more will probably come from near the surface than will be able to escape from deeper in the specimen. The analysis is therefore not uniformly representative of the whole ' interaction volume '. Since we would have to make a very large number of not very accurate assumptions in order to calculate theoretically the volume analysed, this is not normally attempted. The analyst has to be content with the fact that the analysis of a solid specimen is likely to come from a volume about 1 μm $\times$ 1 μm $\times$ 1 μm, which could perhaps be reduced to 0·5 μm $\times$ 0·5 μm $\times$ 0·5 μm at low electron beam energies. Consequently if a feature of interest is smaller than this it cannot be analysed in a solid specimen, although it is possible that some of its elements may be detected. The only way to attempt a quantitative analysis of sub-micron features is with a thin specimen in a TEM or STEM-based analyser.

The minimum limit of detection for any element in a specimen is of particular importance when small concentrations are suspected. It is not of much use to be told " iron cannot be detected in your sample " unless this enables us to put an upper limit on the amount which could be present. The microanalyst, therefore, never gives the answer " there is no iron in your specimen " but instead says " there must be less than (say) 0·1% of iron in your specimen ". To estimate with any accuracy what this minimum detectable limit is in any particular specimen we must look more closely at the X-ray spectrum and the number of ' counts ' in any characteristic X-ray line which is used for analysis. Figure 5·20 (*a*) shows a small part of the complete X-ray spectrum from a specimen—just the single characteristic X-ray line to be used in the analysis and the ' background ' on either side. To continue with the example started previously, let us assume that the line is the K_α line of iron and therefore that the peak corresponds to an energy of 6·40 keV and a wavelength of 0·19 nm (from table 5.1). If this line was detected by an EDS system we would see it displayed as the series of dots, whereas if a WDS system had been used we would only have available the number of counts corresponding to the ' peak ' and ' background ' levels, indicated as P and B respectively in the diagram. In either case, in order to determine the height of the peak (P–B) we need to know accurately the number of counts at both P and B. This is where we meet the first statistical problem: the X-ray counts are arriving at random, not regularly, and statistics tell us that if we make many measurements of the number of randomly occurring events which we detect in a given time we will get a variety of results. There will be a mean value, N, but all the measured values will be scattered about this mean with a standard deviation of $\sqrt{N}$. For example, if X-rays are arriving randomly at an average rate of 100 per second the number we detect in a single second may well be far from 100, but if we make a lot of similar measurements two-thirds of the results will lie in the range $100 \pm \sqrt{100}$, i.e. between 90 and 110. Therefore, if we make a single measurement of a number of X-ray counts we can only quote it as $N \pm \sqrt{N}$; there is a statistical imprecision associated with it.

Consider how this imprecision affects the measurement of P and B. From a pure iron specimen the peak shown in fig. 5.20 (*a*) should be

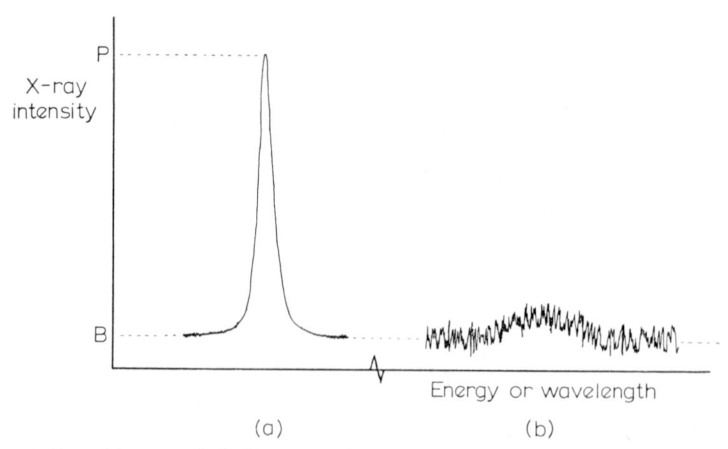

Fig. 5.20. Characteristic X-ray peaks from (*a*) a standard and (*b*) a specimen which contains the minimum detectable amount of the element.

high compared with its background. Perhaps counts have been collected for 100 s and P is 1 000 000 counts. We can, therefore, quote the measurement of P as 1 000 000 ± 1000, that is to an accuracy of about one-tenth of a per cent (1000/1 000 000 × 100). The background count, however, should be much lower, say only 1000 counts; we can therefore only quote it to an accuracy of 1000/1000 × 100 per cent, or about 1%. Because there is this much statistical uncertainty in the background level it appears, as shown in fig. 5.20, as a ‘ fuzzy ’ line. When P is high the imprecision of B does not matter much since it will not greatly affect the value of P–B. However, if we are now looking for a minute trace of iron in a specimen which is predominantly made up of other elements we are looking for the presence of a ‘ peak ’ which is scarcely above the background level, as fig. 5.20 (*b*) indicates. The ‘ fuzziness ’ of the background is now of great concern. It is usually accepted that a ‘ peak ’ can be recognized if it rises above the background level by two or three ‘ standard deviations ’. Let us be optimistic and take the figure of two standard deviations. If the count rate in the background is b counts per second then in t seconds we detect bt counts and the standard deviation is $\sqrt{bt}$. The smallest detectable peak therefore has a P–B of $2\sqrt{bt}$. What concentration of iron in the specimen would give this peak? To answer this we must compare our peak height with the peak height (P–B) given by a standard pure iron specimen (e.g. fig. 5.20 (*a*)). To a first approximation the minimum detectable concentration (MDC) is

$$\text{MDC} = \frac{(\text{P–B})\text{ specimen} \times 100}{(\text{P–B})\text{ pure iron standard}} = \frac{2\sqrt{bt} \times 100}{pt - bt} \quad \text{weight \% Fe,}$$

where p is the peak count rate (counts per second) on the pure iron standard. Simplifying this slightly we get

$$\text{MDC} = \frac{200\sqrt{b}}{(p-b)\sqrt{t}}\%. \qquad (5.2)$$

Clearly the smaller we can make MDC the better. There are only three ways to do this, as equation (5.2) shows; reduce b, increase p or increase t. The easiest of these is to increase the time of analysis and to a limited extent this can be done. Figure 5.21 shows an example of the benefits of this in showing up small peaks. However, it becomes impractical to count for longer than about 1000 s for two reasons—the machine is expensive and often cannot be used for as long as would be needed and, more fundamentally, the electron beam intensity does not stay constant indefinitely, even on the best instruments, and if it varies perceptibly it will no longer be valid to compare the P–B from the specimen with that from the standard, since they cannot be obtained simultaneously.

The other two ways of improving the minimum detectable concentration really depend on the design and functioning of the analysis and detector system. EDS systems tend to be efficient at detecting X-rays and therefore give a high p but also introduce a lot of ' noise ' and hence give a high b. WDS systems are very good at reducing the background (low b) but are less efficient and hence reduce p. It is therefore very difficult to generalize about the minimum detection limit of the EPMA technique. For light elements ($Z<11$) only a WDS can be used and MDC is likely to lie between 0·1% and 1%. For heavier elements it is hotly debated whether EDS or WDS analysers give better results but MDCs should range between 0·1% and better than 0·01%, depending on the element being analysed. This sort of detection limit seems very poor when compared with the results obtainable by many other methods but when one considers the volume which is being analysed the electron probe can be seen in its true perspective. For a not-too-light element we should be able to detect a concentration of 0·01% in a volume of approximately one cubic micrometre (10^{-18} m^3) which is a total mass of only about

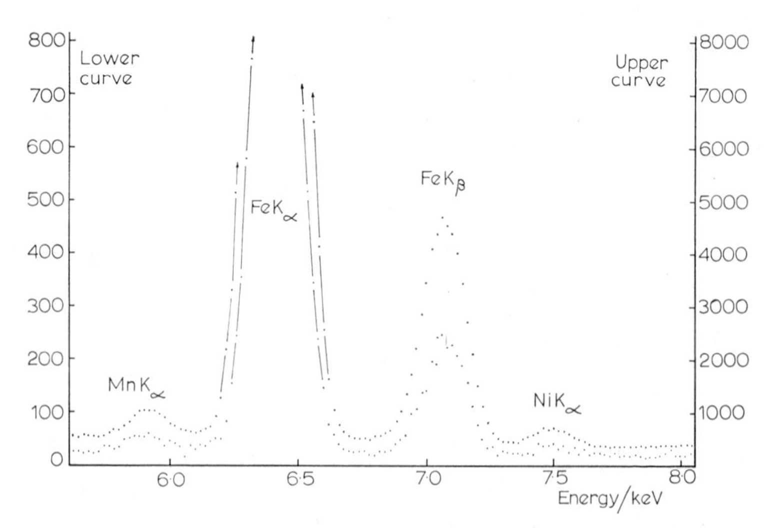

Fig. 5.21. A portion of the X-ray spectrum from a steel specimen. The lower curve shows the spectrum after 10 s collection time whereas the upper curve is the result of collecting for 200 s. The iron K_α and K_β peaks are easily seen in both curves but the presence of manganese K_α and nickel K_α could not be asserted with any confidence from the 10 s curve. After the longer counting time, however, the statistical fluctuation in the background is much reduced and the small peaks can be seen quite easily. Note that the vertical scale is not the same for the two curves.

10^{-15} g. This compares very favourably with chemical analysis ($\sim 10^{-9}$ g at best) and even mass spectroscopy ($\sim 10^{-13}$ g at best); and of these two neither tells us anything about *where* the material was in the specimen. The only instruments to beat the electron probe for sensitivity to small amounts of material are the nose (10^{-18} g on a good day) and the sex attraction organs of the bee (10^{-20} g with a following wind).

Having discussed what we cannot detect in the probe analyser let us turn our attention to the final question for this section: how accurately can we determine the concentration of an element which *is* detectable? The principle of such a quantitative determination is simple and we have already implicitly made use of it while discussing the minimum detection limit. To estimate the amount of an element present we determine the number of characteristic X-ray counts arriving from the specimen in a fixed time interval, N_{spec}, and compare this figure with the number arriving from a standard of known composition in a similar time, N_{std}. The concentration of this element in the specimen, C_{spec}, should then be given by

$$C_{spec} = \frac{N_{spec}}{N_{std}} \times C_{std}, \tag{5.3}$$

where C_{std} is the accurately known concentration of this element in the standard and each value of N is a peak count minus a background count. Now clearly the electron beam cannot be directed at both specimen and the standard simultaneously, so the two counts N_{spec} and N_{std} must be collected at different times. In order to make the comparison implied in equation (5.3) we must be certain that the analysis conditions have not changed between the determination of N_{spec} and the determination of N_{std}. This calls for a great deal of care, both in designing microprobe analysers and in using them to ensure that in particular the electron beam current (i.e. the number of electrons arriving at the specimen per second) remains constant for long periods. In practice this is one of the most difficult things to achieve during probe analysis—ideally the beam current should fluctuate no more than perhaps one-tenth of a per cent while half-a-dozen points on a specimen are analysed and compared with two or three standards. This is likely to take several hours. However, let us assume that by skilled operation of the instrument this problem is overcome and continue to explore the reasons why a delightfully simple expression like equation (5.3) is totally inadequate for the calculation of concentrations.

Most of the complicating factors arise because, inevitably, the specimen is not the same as the standard; in the most usual cases a specimen containing several elements is compared with a series of standards, each of which is a pure element. The specimen is therefore likely to differ from each standard in its density and in the average atomic weight of its constituent atoms. Also the number of characteristic X-rays from the element of interest arriving at the detector per second may be very different. The consequence of these differences is that equation (5.3) may need correcting for some or, in the worst cases, all of the four effects known as *dead time*, *atomic number effect* (Z), *absorption* (A) and *fluorescence* (F). Let us consider the origin of these effects.

The X-ray detector used in the probe analyser, whether it is a proportional counter in a WDS or a silicon detector in an EDS, cannot count a second X-ray for a short interval after the first arrives. There is a similar dead period associated with the counting electronics and the combined effect is to produce a small *dead time* after each X-ray arrives during which any other X-rays are ignored. The inevitable result is that every counter underestimates the number of X-rays arriving per second. The dead time in modern counting systems is so small (generally less than 5 μs) that the number of lost counts is trivial while the X-ray count rate is lower than about 1000 counts per second (c.p.s.). However, at higher rates, which are frequently used, a correction must be applied to allow for this underestimation. An expression such as

$$N_{\text{real}} = \frac{N_{\text{observed}}}{1 - \tau N_{\text{observed}}} \tag{5.4}$$

is easy to use, where N_{observed} is the count rate (c.p.s.) detected and τ is the dead time in seconds. Notice that at 20 000 c.p.s. (a common count rate) with a dead time of 5 μs the real count rate is 10% higher than the observed rate. Clearly this is a most important correction if accurate results are needed. It is particularly important when there is only a small concentration of the element of interest in the specimen. The count rate of characteristic X-rays from the specimen is then likely to be low, whereas the count rate from the standard will probably be very high. The dead-time corrections would then not cancel out in equation (5.3) as they might if N_{spec} was similar to N_{std}. The dead-time correction is, however, the easiest and most accurate correction to apply and gives rise to no problems once τ has been measured for the particular detector.

The *atomic number* effect is rather more difficult to pin down. The mean atomic number of a homogeneous part of a specimen determines (*a*) how far the electrons penetrate before they have lost too much energy to excite further X-rays, i.e. the depth of the 'interaction volume' in figs. 5.2 and 5.16 and (*b*) what proportion of the electrons are backscattered (suffer elastic scattering and leave the specimen) without having the chance to excite any X-rays. These factors affect the number of X-rays generated in a specimen and the distance they have to travel to get out again. When the mean atomic number of the specimen differs considerably from that of the standard, the count rate for an element will not be linearly proportional to the amount present, so equation (5.3) will need correcting. Unfortunately this correction cannot be formulated precisely but approximate methods of varying degrees of sophistication have been developed. As an example of the magnitude of the effect, the correction which must be applied to a measurement of sulphur ($Z=16$) in a stainless steel (mean Z about 27) amounts to 13%. In other words, for the analysis, equation (5.3) needs to be amended to

$$C_{\text{spec}} = 0{\cdot}87 \frac{N_{\text{spec}}}{N_{\text{std}}} \times C_{\text{std}}$$

to take account of the atomic number effect. This is an extreme example—the effect is often much smaller. A measurement of the vanadium concentration ($Z=23$) in the same steel only requires a 1% correction (i.e. a factor of 1·01).

The absorption effect can also be a major source of error. As can be seen from fig. 5.2 many of the X-rays emerging from the specimen will have travelled a considerable distance within the solid and their intensity will have been reduced by absorption. The amount of the absorption depends very strongly on the elements in the specimen, through their mass absorption coefficients, μ in equation (2.6). It is quite likely that the standard and the specimen will have different mean absorption coefficients and therefore, even if X-rays were generated in exactly the same interaction volume in the standard and the specimen, the absorption effects would necessitate a second correction to equation (5.3). The magnitude of this correction can be quite large, especially where soft X-rays (e.g. the K_α lines of light elements, see table 5.1) are emitted from specimens containing heavier elements. A severe example is the measurement of aluminium in a glass containing alumina (Al_2O_3), silica (SiO_2) and lime (CaO). The factor to be incorporated in equation (5.3) just to account for absorption is then 1·86. The absorption correction is often the most significant effect; for example, even the iron analysis in a stainless steel might need an 8% correction (factor = 1·08). The accuracy of any correction procedure depends on the values of μ which are used. Good experimental values are not available for all elements and X-ray lines (particularly for light elements) and this lack of data is a limiting factor in many corrections.

The final effect to be considered is fluorescence. From its very nature (see Section 5.1) fluorescence cannot occur within a pure elemental standard. However, in a specimen containing several elements it must be considered. Fortunately fluorescence is a very 'inefficient' process and only a very small proportion of high energy X-rays excite lower energy fluorescent radiation. However, when elements of nearby atomic number are present, as tends to happen for example in steels (Cr = 24, Mn = 25, Fe = 26, Ni = 28) the fluorescence effect, which gives rise to more of the low energy X-rays than would be expected, can be important. The worst case is usually that of chromium in steels where a correction as large as 15% (factor = 0·85) may be needed.

The three corrections (Z, A and F) tend to be rather complicated to work out and in any case the method of correction generally requires you to know the answer before you start. For instance, you cannot calculate a 'mean atomic number' for your specimen until you know how much of each element is present. Since this information is clearly not available at the start of the analysis it is necessary to use an *iterative* method of calculation: a guess is made of the composition—the corrections are then applied to refine this guess into a 'better guess' and so on until an accurate answer is produced. Because the process of iteration is inevitably tedious, most serious correction of measured data to produce quantitative analyses is carried out using a computer. It is then easy to perform all the required corrections in the appropriate order, which we can simply express, ignoring the mathematics entirely, as the ultimate modification of equation (5.3).

$$C_{\text{spec}} = \frac{N_{\text{spec}} \times (\text{dead time correction}) \times (\text{Z, A and F corrections})}{N_{\text{std}}\ (\text{dead time correction})} \times C_{\text{std}}.$$

The application of computer correction procedures to good results from the analysis of solid specimens using well-characterized standards should enable concentrations to be calculated to about $\pm 2\%$ with a WDS and $\pm 6\%$ with an EDS. However, these results can only be obtained by extremely careful experimentation and where there are large amounts of the element present (say about 10%). Thus, although it might be possible to measure the concentration of copper in a brass as $70.0 \pm 1{\cdot}4\%$, the carbon content of a steel could scarcely be calculated and quoted better than $0{\cdot}3 \pm 0{\cdot}2\%$ because of the low concentration and the lack of accurate values for absorption coefficients of the 'soft' carbon K_α X-rays.

One of the most important parameters in the correction equations is the X-ray take-off angle (ϕ in fig. 5.6) which determines the X-ray path lengths in the specimen. It is extremely important that this should be the same on both specimen and standard and, indeed, at all points of the specimen which are to be analysed. Consequently any specimen for quantitative analysis must be flat on the scale of the electron beam diameter. It is usual to polish specimens immediately before analysis with an abrasive of 1 μm or $\frac{1}{4}$ μm particle size so that only very fine undulations are present. A flat specimen is also virtually essential for any analysis using a WDS since, as fig. 5.6 shows, the specimen must lie exactly on the Rowland circle for efficient detection of the X-rays. A rough specimen could not be moved sideways at all without re-aligning it on this circle which would be a great inconvenience. It is one of the apparently great advantages of the semiconductor detector (EDS) that it is extremely insensitive to the position of the specimen and, therefore, rough specimens can easily be ' analysed '. However, if a quantitative analysis is required the take-off angle must be known and, therefore, a polished specimen is again necessary. It is worth bearing in mind that even a qualitative analysis from a point on the specimen using an EDS can be misleading because of the effects shown in fig. 5.13.

Although the preparation of the specimen is a key factor in quantitative analysis there are also problems associated with the standards. For many elements, particularly the metals, it is extremely easy to prepare a polished sample of the pure element. However, many of the other useful elements cannot be used pure since they are normally gases, liquids or very reactive solids. Oxygen, nitrogen, chlorine, mercury and sodium are a few of the more common, and standards for the analysis of these elements must therefore be compounds. Few compounds are really suitable since they must be of accurately known homogeneous composition in the solid form and be stable under electron bombardment. For instance, common salt would seem to be an excellent standard for sodium and chlorine but it is deliquescent and therefore a block of rock salt must be freshly cleaved just before it is put into the vacuum chamber of the analyser, otherwise the dissolved water would alter the composition of the surface layers in an unknown way. Again, alumina (Al_2O_3) could seem to be an appropriate standard for oxygen—it is very stable and readily available. However, many solid forms of alumina are porous and are therefore less dense (and contain fewer oxygen atoms per unit volume) than their formula would suggest. Many pitfalls such as these have been recognized and circumvented but there are still some elements for which it is extremely difficult to find a satisfactory standard.

Since the major emphasis in this section has been on the quantitative analysis of metallic or mineral specimens, which by and large are stiff solids which can be polished flat without damage, it may be useful to summarize some of the problems involved in applying these techniques to biological materials. The majority of biological samples are unsuitable for accurate quantitative analysis because they do not fulfil three of the important criteria described as essential earlier in this section:

(*a*) They are rarely stable under electron bombardment, frequently changing their volume and their chemical nature;

(*b*) they cannot normally be considered homogeneous in composition over the volume being analysed, which is often large (more than 10 $(\mu m)^3$) because of the low atomic weight of most of their constituents; and

(*c*) they do not normally maintain a flat surface (and hence a constant take-off angle) inside the microscope. Even if the original specimen has a flat, smooth surface it is unlikely to remain so under the dual effects of high vacuum (causing dehydration and collapse) and electron bombardment (causing chemical changes and hence often collapse).

When to these basic difficulties are added the problems associated with analysing light elements and our lack of knowledge of the absorption coefficients of even heavy element (hard) X-rays in light elements it can be seen that the quantitative analysis of most biological matter is out of the question. However, so much new information is still to be gathered, even on a qualitative basis, from biological cultures and sections that the various forms of probe analyser will undoubtedly continue to be used extensively in these fields.

5.6. *Some applications of electron probe microanalysis*

Four applications have been selected which, although they represent only a few of the fields in which EPMA has proved useful, illustrate most of the analytical techniques which have been discussed in this chapter.

5.6.1. *Corrosion of stainless steel*

The ash which results when fuel oil is burnt is rather corrosive. Even ‘ stainless ’ steel parts in the burner are subject to corrosion. The rate of attack is dependent on the exact composition of the fuel ash, which in turn depends on the composition of the oil, and hence on which oilfield the supply came from. In order to ascertain which components of the oil are most active in promoting corrosion a laboratory investigation has been carried out in which the reaction between various ashes and steels has been simulated.

The example shown here (fig. 5.22) is of a synthetic ash containing 30% sodium sulphate and 70% vanadium pentoxide (both common constituents of real fuel ashes) which has been allowed to attack a 9·3% nickel-containing steel for 100 hours at 675°C. A polished section of the ash/steel interface was then examined in a combined

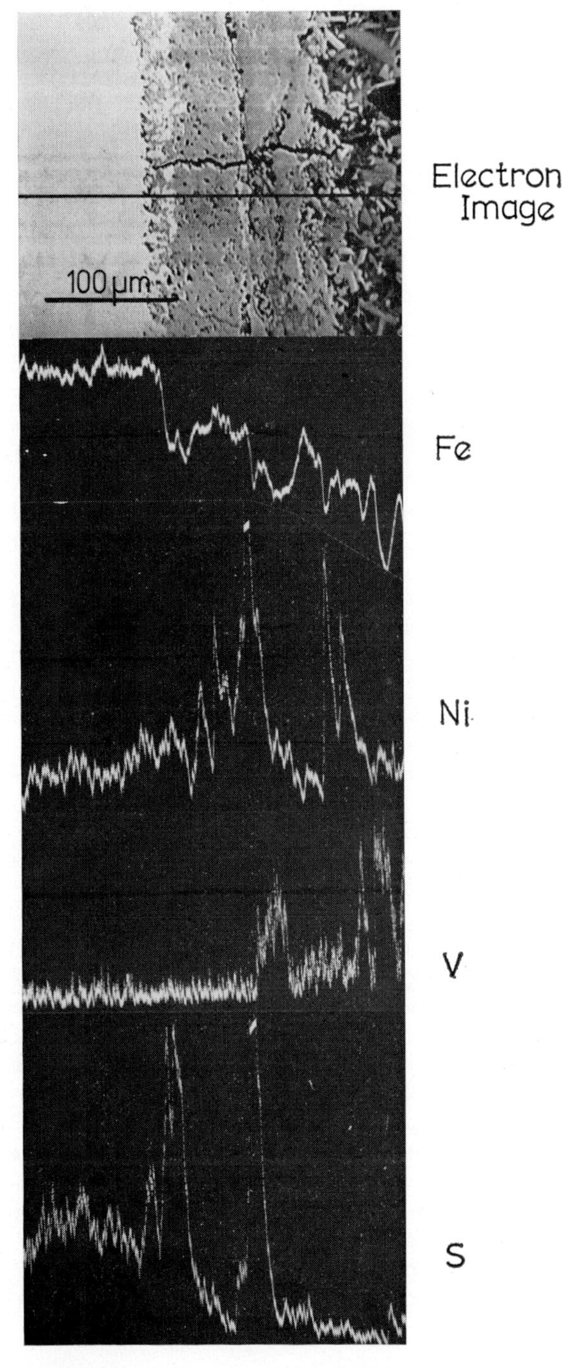

Fig. 5.22. The traces of four elements across the interface between a slag and the steel substrate. Each of the traces was taken from the line indicated. (D. Parkin.)

SEM/EPMA. The micrograph illustrated in fig. 5.22 shows clearly the severe attack of the originally flat metal surface and indicates that two distinct layers have formed between the ash (on the far right) and the metal. Using a crystal spectrometer (WDS) the distribution of each of four elements was displayed on the c.r.t. in turn. Each distribution was taken from the same line, indicated on the micrograph, so that if particular elements occurred together (i.e. in a compound) a peak appeared at the same point on each trace. The distribution of iron, nickel, vanadium and sulphur are shown in the figure and various conclusions can be drawn: clearly the vanadium has not penetrated to the inner of the two layers, which contains high concentrations of nickel compared with the original alloy. A further conclusion is that the sulphur appears to be associated with iron peaks and hence the formation of an iron sulphide or sulphate seems likely.

From a large number of observations such as these, on systems with different metal and ash compositions, it has been possible to discover what chemical compounds and metallic phases are formed during corrosion and hence what the thermodynamic behaviour of the system must be. It is then possible to recommend which constituents of the original fuel are likely to be most corrosive on any particular steel, and hence to ensure that the most appropriate steel is used in each case.

5.6.2. *Two biological applications of EMMA*

The two applications to be described here are fairly typical of biological uses of analytical microscopy in that they are qualitative analyses. However, despite the virtual impossibility of performing a quantitative analysis, as discussed in Section 5.5, the qualitative information is tremendously useful.

Figure 5.23 is a micrograph of a section of a human cornea. The patient was suspected to be suffering from Wilson's disease, one of the symptoms of which is the appearance of a brown ring in the cornea. The disease is caused by absorption of copper from the alimentary tract and its subsequent deposition in the eye, brain and liver. In order to diagnose the disease positively it was necessary to identify the deposits in the cornea. The small dark particles appearing in Descemet's membrane (see fig. 5.23) are only $0{\cdot}2\,\mu m$ or less in diameter and EMMA provided virtually the only way of identifying them. Using an LiF crystal in the WDS of an EMMA-4 instrument, copper was detected from each of several of the particles, confirming the diagnosis.

A second example of an EMMA analysis is shown in fig. 5.24. This shows a single human sperm cell and the X-ray spectrum from its head, taken using a semiconductor detector (EDS). The sperm head clearly contains many elements, as the labelling of fig. 5.24 (*b*) indicates (the gold peak is in fact spurious since it arose from the gold grid on which the specimen was supported). The elements of greatest interest were copper and zinc: treatment of sperm with copper causes them to lose their ability to ' swim ' and also brings about a drop in the zinc concentration of the head, possibly because of the inhibition of an enzyme.

Clearly, although quantitative estimation of the amounts of copper and zinc present is impossible, it is quite feasible to estimate the ratio of the heights of the copper and zinc peaks or the ratio of both to some other element such as phosphorus, chlorine or potassium. Analytical electron microscopy is therefore a powerful technique for studying the variation of copper and zinc levels in a large number of sperm cells and thus gaining a further insight into fertility.

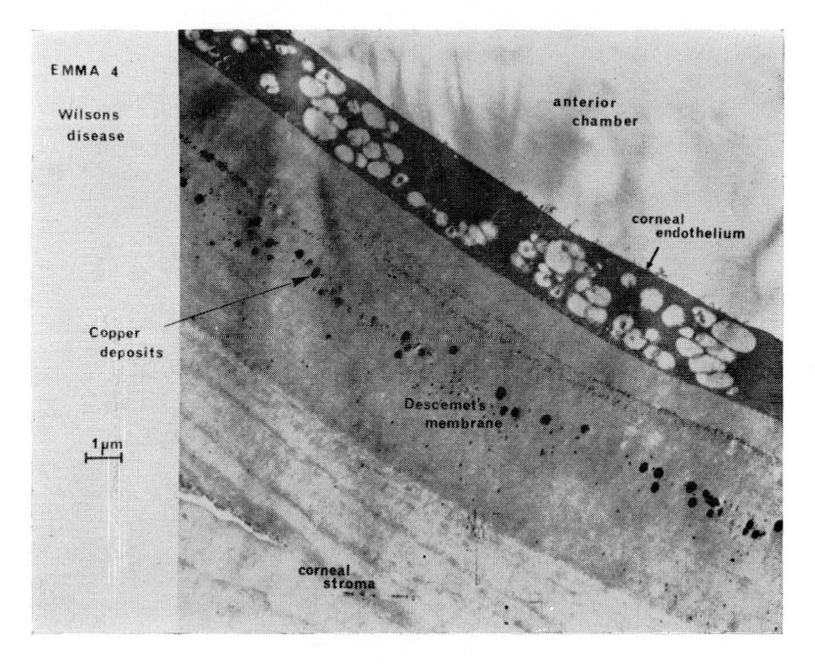

Fig. 5.23. Copper deposits in the cornea of a sufferer from Wilson's disease. Each particle could be analysed using EMMA-4. (J. A. Chandler.)

5.6.3. *The microprobe analysis of glass*

Glasses are a fascinating range of materials to analyse. They may contain a large number of elements since, as well as the almost essential silica (SiO_2), common additions are lime (CaO), alumina (Al_2O_3), soda (Na_2O), potash (K_2O) and less frequently compounds of

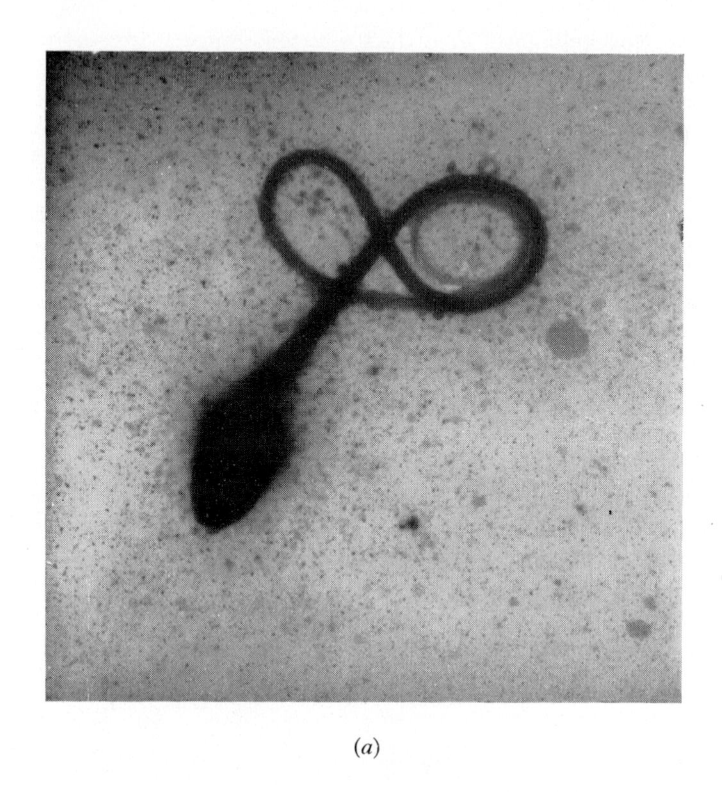

(*a*)

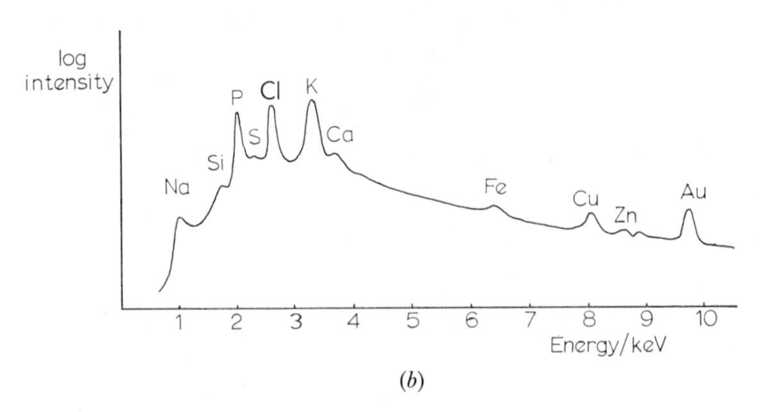

(*b*)

Fig. 5.24. (*a*) A single human sperm cell, dried down out of solution for analysis. (*b*) The X-ray spectrum collected from a point within the head of the sperm using EMMA-4. (J. A. Chandler.)

lead, barium, magnesium or transition metals. Very many types of glass, with distinctly different composition, are in use and non-destructive analysis of small fragments is often undertaken for forensic purposes, or to locate the country of origin of archaeological glass specimens, as well as for more straightforward scientific reasons.

However, glass specimens are not easy to examine in a conventional EPMA. Most glasses are non-conductors and contain the very light element oxygen and a number of other fairly light elements such as sodium, aluminium and silicon. In addition it is frequently found that the apparent concentration of the alkali metals sodium and potassium changes while a glass specimen is being analysed. The reason for this behaviour is not hard to find, but it has proved difficult to prevent. Sodium, which is subject to the most serious changes, is added to glasses in order to soften them so that they can be shaped at lower temperatures. The softening occurs largely because sodium ions in the glass diffuse very rapidly through the structure; however, it is this readiness to diffuse which presents the microanalyst with problems.

During microprobe analysis of glasses electrons are being deposited below the surface of a non-conducting material, and at the same time are heating up the 'interaction volume'. The electrons, and their negative charge, normally come to rest towards the bottom of the interaction volume (fig. 5.2), while the sodium X-rays (soft) only escape from quite near the specimen surface. Consequently as the analysis continues an increasing negative charge is built up below the 'warm' region which is being analysed (fig. 5.25 (*a*)). The result is that the positively charged sodium ions tend to diffuse towards the negative electrons and the volume being analysed becomes depleted in sodium. The result is that very frequently the sodium count rate drops as the analysis proceeds, as illustrated in fig. 5.25 (*b*).

Clearly it is hopeless to try to deduce a quantitative analysis, which may require counting times of 100 to 1000 s from a specimen whose composition is varying so rapidly. The only ways round this problem are

(*a*) to deposit fewer electrons in the glass or to deposit them where they have less influence,
(*b*) to reduce the heating-up caused by the electron beam and/or
(*c*) to complete the analysis before the sodium ions have time to move appreciably.

Using a crystal spectrometer it is necessary to operate with such high electron beam currents (10^{-6}–10^{-8} A) that (*a*) and (*b*) are impracticable and the only solution is to move the specimen so that the beam does not remain long on a single spot. This of course removes one of the great advantages of microprobe analysis and implies that we cannot analyse a point in a glass specimen. However, because the semiconductor detector (EDS) is so much more efficient at collecting X-rays, much lower beam currents can be used (10^{-9}–10^{-10} A). This does achieve points (*a*) and (*b*) and, with care, glasses can now be analysed for sodium using an EDS system. This is a good example of the advantages to be gained by using an EDS system but it does emphasize that it is necessary to appreciate what is happening inside the specimen before a sensible solution to an analytical problem can be suggested.

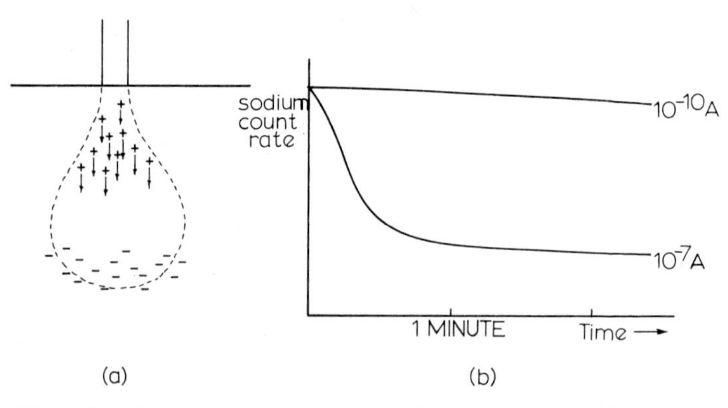

Fig. 5.25. The change of composition within the analysed volume of a glass specimen. (*a*) Negatively charged electrons collect below the surface and attract the positive sodium ions away from the surface. (*b*) At high electron beam currents (10^{-7} A) the sodium count rate drops within a minute to the background level; if the beam current is reduced to 10^{-10} A the decay is very much reduced and an analysis can be attempted.

5.6.4. *The reaction between iron and zinc*

The reactions which occur between solid iron and liquid zinc are of great practical importance in the widely used process of 'hot-dip galvanizing'. Steel sheet is coated with a thin protective layer of zinc by being passed at a controlled rate through a bath of molten zinc. Clearly for maximum economy the process should result in the formation of a uniform coating which is as thin as it can be while still affording sufficient corrosion protection.

The equilibrium diagram of the iron–zinc system indicates that five intermetallic phases should be formed between pure iron and pure zinc. In the galvanizing of plates these should be present as thin layers between the steel base and the external zinc layer. The microprobe analyser provides a good way of determining the thickness of each layer, since the iron concentration should drop suddenly at each interface. Figure 5.26 shows an example of iron and zinc X-ray traces across one such specimen. The successive layers should be, from left to right, pure zinc, zeta phase with composition 6.9% Fe, delta phase, a mixture of delta and gamma phases, a mixture of gamma and alpha phases and finally pure iron. The traces shown in fig. 5.26 bear out this explanation except for two features: between the zeta and delta layers there appears to be a narrow layer which is

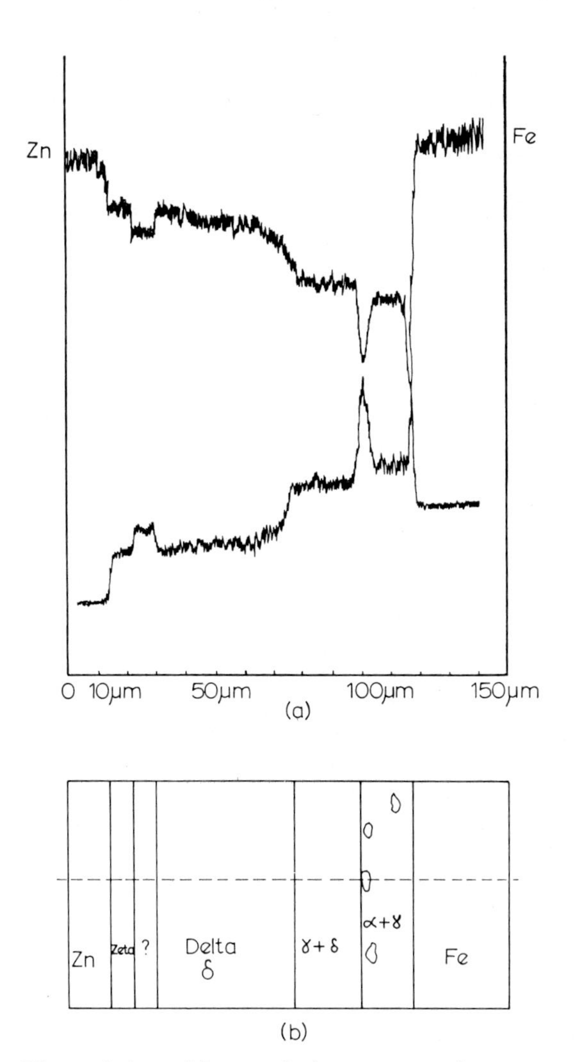

Fig. 5.26. The variation of iron and zinc concentration across the various layers of an iron–zinc interface. (N. Short.)

slightly richer in iron than either of its neighbours and between the $\gamma + \delta$ and $\alpha + \gamma$ regions there is an even narrower layer which is very iron-rich. These two layers were analysed in more detail by placing the stationary electron beam on each in turn and performing a quantitative analysis against pure iron and zinc standards. The very iron-rich region proved to have the composition of the α-phase and examination of the SEM image disclosed that this region was not a complete layer but only ' islands ' of trapped α-phase as indicated in fig. 5.26 (*b*). The analysis of the unknown layer which was nearer the zinc proved less easy to understand. Its composition was 20% Fe, which is quite a long way from zeta phase (6·9% Fe) and delta phase (10% Fe). This composition does not correspond to any known phase in the iron–zinc equilibrium phase diagram and must therefore either be a frozen non-equilibrium phase, an unreported equilibrium phase or, perhaps most likely, a layer of trapped oxide. Work is still continuing on this problem, which has provided us with a classic example of the EPMA technique, throwing up unexpected information during what was thought to be a fairly mundane and routine investigation.

CHAPTER 6

comparison of electron microscope methods with other techniques

ELECTRON microscopy and analysis have been rapidly changing fields even in the short period of 25 years since the instruments became widely available. The development of some of the other techniques which can be used to provide similar types of information has been equally dynamic. It is therefore very difficult either to predict future developments in electron microscopy or to make an accurate comparison with the currently available competing techniques. All that this final chapter can do is to outline some of the more obvious directions in which the subject is moving and to draw general comparisons with some other techniques, paying particular attention to the unique advantages and disadvantages of each. To do this it seems most convenient to split the subject into the three distinct areas in which electron beam techniques excel: ' seeing ' small features and the relationships between them; analysing the structure of materials on a fine scale; and determining the composition of very small selected regions of a specimen.

6.1. *Microscopy*

We considered in Chapter 1 the limitations of the light microscope and how the two main types of electron microscope could overcome some of them. After immersing ourselves in electron beam techniques for four chapters it is now time to re-assess the relative merits of the two types of technique and also to ask the question " What more would we like to be able to see, and how can we do it? "

Let us first consider the internal microstructure of materials. Whether our specimen is a biological tissue or a structural alloy the points of interest are the size, shape and relationships between microstructural features—cells, nuclei, viruses, grains, dislocations, precipitates or whatever they may be—which cannot be deduced from the external morphology of the material. In both biological and materials sciences the light microscope has been an invaluable tool for investigating this internal structure. The biologist has often been able to cut his material into transparent slices and see the microstructural arrangement directly. The materials scientist, on the other hand, has had to deduce information about internal structure from the appearance of a flat etched section of the solid. In both cases, however, many years of experience and development have meant that nowadays a very great deal can be learnt from light microscope techniques. Not the least of the advantages of the 'simple' techniques is that colour can be seen. Without this information many a metallurgist or mineralogist would be struggling to recognize even the commonest structures. Another advantage, this time mainly in the biologist's favour, is that the specimen for light microscopy can be fairly thick and does not have to suffer collapse or drying in a vacuum. Again, in the light microscope it is easy to relate the structure you see to the scale of the whole specimen; low magnifications can be used to bridge the gap between high magnification micrographs and what the eye can recognize. Finally, all these advantages can be had for somewhere between one-tenth and one-hundredth of the cost of a transmission electron microscope.

Chapter 3 should have made obvious the unique advantages of the TEM; the materials scientist is for the first time able to see directly inside a solid; dislocations, stacking faults and small precipitates become identifiable features rather than theoretical concepts; the biologist is able to resolve smaller details—he (or she) can at last see ribosomes, the structure within mitochondria (the chrystae), and the fine layered structure of the membranes surrounding cells and cell nuclei. Information can be deduced about these features by other methods but almost without exception they are indirect methods which give data about the *average* behaviour of the microstructure. For instance, metallurgists have for many years been using X-ray diffraction methods to learn something about crystal structures, crystal defects and grain orientations; however, this technique tells us very little about the structure of a single grain, about the interactions between individual lattice defects or about the misorientation between one grain and the next. All these factors are essential to an understanding of metal deformation processes, and they can only be studied by TEM.

What more would it be desirable to see? The two most likely answers to this question, from both the biologist and the materials scientist, would be " a thicker section of the specimen " and " individual atoms and molecules ". From a thicker specimen one can learn more about the three-dimensional relationship between the various features which are only usually seen in section—for instance, the real shape of the mitochondria which generally appear as a bundle of disconnected sausage-shaped outlines. The other great advantage of a thicker section is that the structure which is seen within it is more likely to be representative of the bulk of the specimen and less likely to be influenced by the proximity of the two surfaces of the section. This is of particular importance to the metallurgist who knows that dislocations and other crystal defects can be attracted to a surface and may be annihilated when they reach it. The structure of a very thin foil may therefore bear little resemblance to the structure far from the surface of a bulk specimen.

There are several ways in which it is possible to see through thicker sections, and all of them are being actively developed at the moment. The two most promising approaches are to raise the accelerating potential of the electron beam, perhaps from 100 kV to 1 MV or more, and to use a scanning transmission system (STEM). The use of higher voltages undoubtedly allows thicker specimens to be examined; however, a ten-fold increase in voltage at present costs ten times as much but only gives rise to a four or five-fold increase in usable specimen thickness. Also, a million volt microscope needs to be about as high as an average house so it is hardly surprising that there are less than a dozen in the U.K. Nonetheless high voltage electron microscopy (HVEM) has already proved extremely useful in studying materials composed of very heavy elements, which could scarcely be studied at 'ordinary' voltages, and in obtaining three-dimensional information from specimens which are relatively light (for example, biological tissue). One tremendous possibility is that living cells could be studied, since a beam of million volt electrons should be able to penetrate a 'window' and a cell in its wet state, surrounded by air. It would obviously be very exciting to be able to study tissue without having to dry and evacuate it and therefore to see the true undistorted arrangement of all components, or possibly even the movement of small features.

The use of STEM techniques to increase the usable thickness of specimens is currently receiving a lot of attention because it is possible to modify existing transmission or scanning electron microscopes so that an improvement of perhaps two or three times the specimen thickness can be obtained without the need for an increased accelerating voltage. The fundamental reason for this improvement is, as stated in Chapter 4, that almost all the electrons which pass through the specimen can be collected and may contribute to a STEM image, whereas in the TEM the specimen must be so thin that the electrons are not significantly slowed down by their passage through it; otherwise chromatic aberrations in the image-forming lenses become intolerable. It is an extremely simple matter to fit a detector for transmitted electrons to convert an ordinary SEM to a STEM. However, because of the low voltages (<30 kV) conventionally used in SEM's this only brings the penetration of the SEM up to that of a common TEM (using ~ 100 kV). If a scanning system is fitted to a conventional 100 kV TEM, however, a really worthwhile improvement in useful specimen thickness is realized. This sort of modification is very likely to spread because it is cheaper than buying a 300–500 kV TEM and in addition offers all the advantages of the scanning principle of image formation.

The two approaches just outlined appear to offer an encouraging prospect to the microscopist who needs to see through a thicker section of his material. What prospects can we offer in the other direction mentioned above, towards the imaging of individual atoms or molecules? The answers here must be different for the biologist and for the materials scientist. Transmission electron microscopes are already capable of sufficient resolution to image many biochemical molecules. The reasons why we do not in general see such molecules are inherent in the specimen not in the microscope. The limitations are those of *contrast* and *specimen stability*. Because biological tissue shows virtually no contrast and is rapidly destroyed by the electron beam it is normally stained with heavy atoms before observation. We are therefore only looking at a skeleton of heavy atoms, the resolution of which is normally considered to be 1–2 nm. Much molecular detail is therefore lost to view, even though the microscope may be capable of resolving details separated by only 0·2 nm. The most hopeful approach to this problem appears again to be via a STEM technique. It is possible to achieve a resolution with a STEM instrument which approaches that of the TEM. It is necessary for these purposes to use a field emission gun (see Section 2.3) but the high vacuum technology which is then essential is rapidly becoming more common. The promise of STEM as applied to biology is that by analysing the energy of the electrons after they have passed through the specimen but before they enter the detector the image contrast can be made quite sensitive to atomic number. Eventually it should therefore be possible to show the arrangement of individual atoms (heavier than carbon) in unstained molecules. We can hope that the thorium atoms shown in fig. 2.11 are only the first rough results of what will be a delicate and sensitive technique for studying large molecules.

The materials scientist is just as interested as the biologist in the prospect of ‘ seeing ’ individual atoms, ions or molecules but it does not seem that electron microscopy is going to be the most appropriate technique. The problem is that most metallic and ceramic materials are homogeneous on a very fine scale; that is, each atom is surrounded by similar atoms. The features which would be of interest on this scale are vacancies (missing atoms), substitutional or interestitial impurity atoms, or the exact arrangement of atoms at, for instance, grain boundaries. However, none of the contrast mechanisms discussed in Chapter 3 is capable of showing up these defects directly in either crystalline or amorphous specimens. A more suitable technique for making defects visible on the atomic level is the field ion microscope (FIM). In this technique a high potential is applied between a fine-pointed specimen of tip radius r and a screen at a distance R (fig. 6.1 (*a*)). In the intervening space a low gas pressure is maintained: gas atoms are ionized in the very high field near the surface of the specimen. Each ion formed is then attracted directly to the screen, where it contributes a speck of light to the image. If the specimen is cooled the ionizations occur preferentially just above atoms in the specimen and the image on the fluorescent screen consists of a pattern of bright spots corresponding to individual atoms in the specimen. An example of an FIM micrograph is shown in fig. 6.1 (*b*).

The magnification of the FIM image is simply R/r and hence in order to make atoms visible (i.e. make their image bigger than 0·1 mm) r, the radius of the tip of the specimen, must be less than 1 μm. Consequently although images such as fig. 6.1 (*b*) can be used to study single vacancies and grain boundary or dislocation structures in terms of atomic arrangements, only a minute region of the specimen can be studied. Not only is the method restricted to electrically conducting materials (so that the potential can be applied) but the difficulties of ensuring that the structures observed are representative of the bulk material are even more formidable than in the case of electron microscopy. Nevertheless the FIM offers one of the only ways of extending the resolution available in the TEM down to truly atomic dimensions.

Is there a chance of improving the scanning electron microscope so that it can give a resolution of surface detail equivalent to that obtained on internal structure by the TEM and STEM? The answer would appear to be ' no ', since inevitably a solid specimen will scatter the electrons in the electron beam in a fashion similar to that already shown in fig. 5.2 and hence any secondary effect (electrons, X-rays, light, etc.) will be produced from a region larger than the electron beam diameter. Thus even if a beam as fine as 0·1 nm is produced from a field emission gun it will only be able to resolve very fine detail if the specimen is only a few atoms thick, as is the case for high resolution STEM specimens. Undoubtedly the routinely achievable resolution of ' ordinary ' SEMs will continue to improve but it seems unlikely that a resolution better than a few nanometres can be obtained from a solid specimen.

The most important advances in scanning electron microscopy are likely to be in other directions than the quest for resolution.It is the experience of most microscopists that the vast majority of studies do not require very good resolution, but are more concerned with the depth of field available with the instrument. The developments which are now technically possible and which are likely to spread include stereo imaging, quantitative image processing and in-microscope experiments. Many of these techniques are inter-related: for

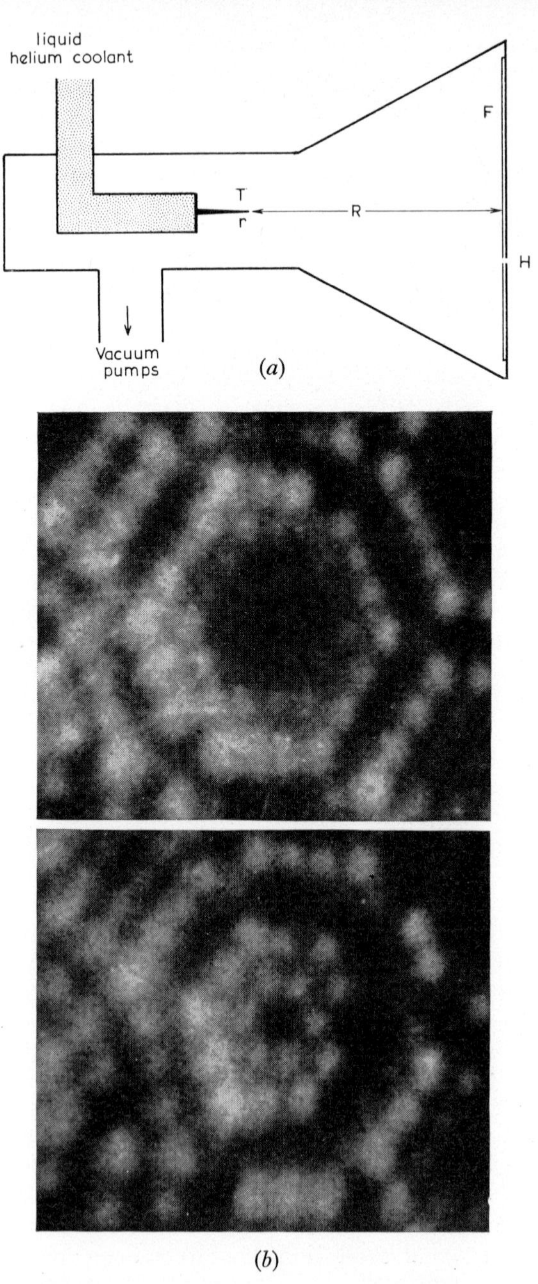

Fig. 6.1. The schematic layout of a field ion microscope. A high voltage is applied between the fluorescent screen (F) and the fine tip of the specimen (T). The image magnification is given simply by R/r. A hole (H) is shown, through which ions could be extracted for analysis when the instrument is used as an ' atom probe '. (*b*) A pair of field ion micrographs of a quenched platinum specimen. A vacancy (vacant atomic site) can be seen in the centre of the lower micrograph.

example, if you wish to cut, scratch or prod your specimen while it is in the microscope then it becomes essential that you can view it in three dimensions. Consequently stereo viewing has been developed to the stage where it is possible to view two TV screens through a pair of binoculars and see the image in 3-D while the specimen is manipulated inside the chamber of the SEM.

It is becoming increasingly common for experiments involving the specimen to be carried out in the microscope. Thus already it is a fairly routine matter to deform the specimen, perhaps fracturing it either by bending or by tensile strain, while recording the image on video tape or cine film. Other ' dynamic ' experiments have involved oxidation of heated specimens (the small amount of gas which is necessary can be pumped away continuously by the vacuum pumps) and the examination of microscopic organisms in a drop of water. It is very likely that the next ten years will see a great increase in the number of this sort of experiment; as the two examples just quoted suggest, it is even possible to get round many of the apparently insuperable disadvantages of having to work in a vacuum. As a rider to the last few paragraphs it might be emphasized that it is not easy to match the vacuum requirements of all the types of SEM in one instrument. It would be unwise at the moment to predict the development of an instrument capable of 10^{-10} Torr at the electron gun (for field emission) *and* 10^{-2} Torr in the specimen chamber (for oxidation or ' wet ' experiments). We shall probably have to choose whether our problem needs the help of an ultra-high vacuum, high resolution SEM or STEM, or would be better served by a low resolution, robust SEM with a vast specimen chamber full of experimental equipment. The contrast is rather like that between a Grand Prix racing car and a ten-ton lorry, with no disrespect intended to either.

A final class of technique deserves a mention before we leave the subject of microscopy. A great deal of work has gone on into ways of improving the clarity of electron micrographs *after* they have been taken. It is possible by several techniques to remove some of the unwanted ' messy ' background of a picture (noise) while retaining or enhancing the features which are to be studied. The most familiar example of this is the computer processing of television pictures from the Apollo moon landings, which are nowadays so much better than the earlier results. Virtually all the moon pictures now shown on television have been processed by an enormous computer at Houston. However, there are other less elaborate ways of achieving similar effects with electron micrographs. Many of these methods make use of optical diffraction and are particularly suited to enhancing periodic detail. A striking example of what can be achieved is shown in fig. 6.2. Although there are pitfalls in this sort of image manipulation, since to a certain extent one can only enhance what one

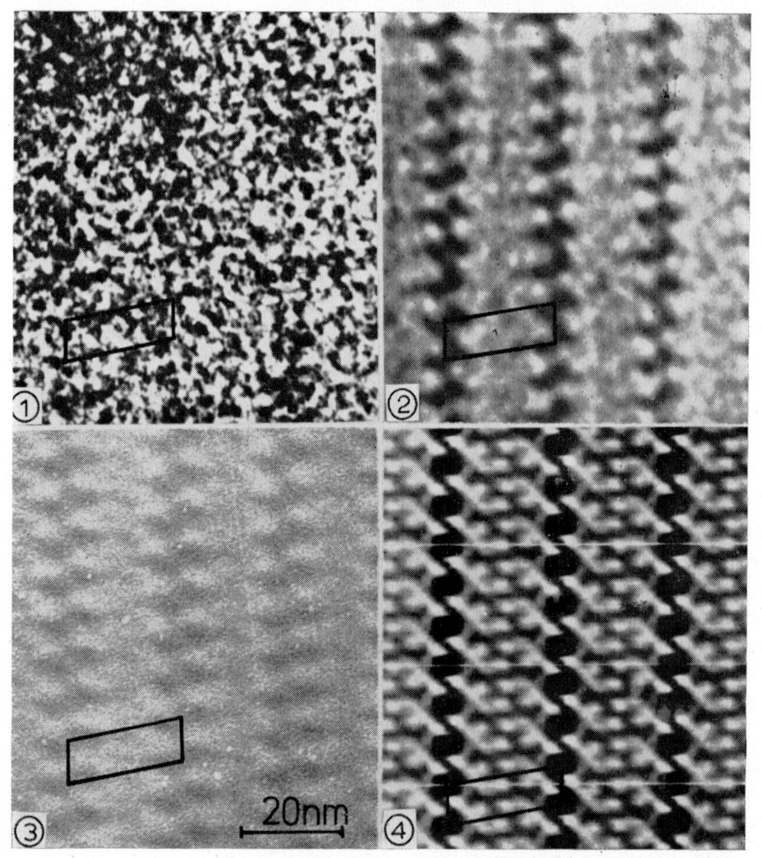

Fig. 6.2. The effect of various types of image processing on an electron micrograph. The micrograph at the upper left is the original, taken from a negatively stained cell wall of the green alga *Chlorogonium elongatum* The second and third micrographs are the result of optical processing techniques and show clearly that a regular structure is present. The fourth micrograph (bottom right) has been reconstructed by a computer from the optical diffraction pattern of the original. Very much more detail is visible. The same region, apparently the unit cell of crystalline structure, has been outlined in each micrograph. (K. Roberts.)

suspects to be there, it will undoubtedly be very helpful in for instance interpreting the three-dimensional arrangement of molecules in small biological organelles.

6.2. *Structure analysis*

Electron diffraction has long been a major technique for structural analysis. We have seen in Chapter 3 how it can be carried out very easily in a transmission electron microscope and in Chapter 4 the extension of similar techniques to scanning microscopy (both SEM and STEM) was described. In principle, therefore, we can obtain information about the crystal structure, lattice parameters and orientation of both solid and thin specimens. The improvements which are likely to be forthcoming in the next few years are mainly concerned with reducing the size of the selected area of the specimen from which the diffraction pattern (transmission pattern or ECP) can be taken. Thus while at the moment it is not easily possible to obtain a pattern from a region smaller than 1 μm in diameter, we can expect this to be reduced in the future. Already the use of higher accelerating voltages in the TEM has meant that smaller areas can be selected and even greater advantages are to be expected from the STEM technique. This will enable particles at least an order of magnitude smaller than at present to be structurally analysed by electron diffraction, and optimistic estimates are already being made that regions as small as 10 nm across will be studied in this way.

There are many other diffraction techniques in regular use for the study of microstructures, most of them employing X-rays rather than electrons. Let us consider what are the particular advantages and disadvantages of electron diffraction carried out in an EM and what additional information some of the other methods might give.

The main advantage of electron microscopic diffraction methods is the one which has been emphasized throughout this book. The information, in this case a diffraction pattern, can be taken from a *small, identifiable* region. In many respects more *accurate* information can be obtained by X-ray methods, but without exception these methods cannot indicate exactly where the analysed region is within the specimen. For several reasons X-ray diffraction is potentially (and in practice) a more accurate method of determining lattice parameters than electron diffraction. X-ray lines are often sharper and, because of the longer wavelength, they are scattered through larger angles; both these factors make it easier to measure Bragg angles for X-rays more accurately. In addition calibration of the camera length (L in fig. 2.12) is much more difficult in the electron microscope since a small change in position of the specimen is magnified by the objective lens into a relatively large change in camera length. A further advantage of X-ray diffraction is that since a large volume of the sample is irradiated a powdered specimen can be used and all possible orientations of the crystal can contribute to the diffraction pattern. In contrast, in the EM only a single orientation is likely to be within the field of view at one time and, therefore, many patterns of different grains, or the same grain in different orientations, must be taken before all the possible diffraction information has been collected. On the other hand, X-ray diffraction is incapable of identifying a *particular* small feature and is also rather insensitive to small quantities of the selected phase. For instance, if we wish to identify a constituent of an alloy which is only present as about 1% of the solid we can reasonably hope to find the unknown phase in the electron microscope and identify a single region of it; we would, however, be unlikely even to detect its presence by most X-ray diffraction techniques, let alone identify it. However, if we then had to decide which of two different phases, with lattice parameters 0·201 and 0·202 nm, was the one being examined we would require accuracy beyond the scope of the EM and the only approach would be to isolate the unknown phase and carry out X-ray diffraction. This might well prove difficult and also would certainly preclude any possibility of determining the orientation relationship between the unknown phase and the matrix. From this type of discussion it is easy to see that there is a place for both techniques and there are many problems which can only be solved using both in collaboration.

The vast difference in the volume of material analysed by X-ray and electron diffraction (cubic millimetres against cubic micrometres, or 10^{-9} m^3 against 10^{-18} m^3) may be an advantage in favour of electron diffraction where small particles have to be analysed but as usual with microscopic techniques brings the problem of whether the structure observed is typical of the bulk material. For some problems it is necessary to study a small volume of a very different shape: in the study of the structure of surface layers even the small penetration of 100 kV electrons is too much and the use of X-rays is quite out of the question. Two other techniques using electrons are then useful. The first, which can often be carried out in a slightly modified TEM, is reflection high energy electron diffraction (RHEED). If diffraction is made to occur from the surface layers of a solid specimen, then the effective penetration of even 100 kV electrons (' high energy ' in this context) is very much reduced. Figure 6.3 makes this clear. Since d, the effective penetration, is related to t, the actual penetration, by

$$d = t \sin \theta_B$$

and θ_B is typically less than 1° we can effectively take a reflection diffraction pattern from a region 50 to 100 times thinner than that analysed by a transmission pattern. In order to obtain a reasonably sharp pattern we generally need to take the reflection pattern from a rather larger area of the specimen than in the transmission case, but this is often not a particular disadvantage. It is also quite difficult to align the very flat specimen so that the electron beam hits it at such a small angle, but this is merely a practical problem and does not invalidate the use of the technique. RHEED patterns are analysed in the same way as normal transmission patterns, although because of the presence of the solid specimen only half a pattern appears.

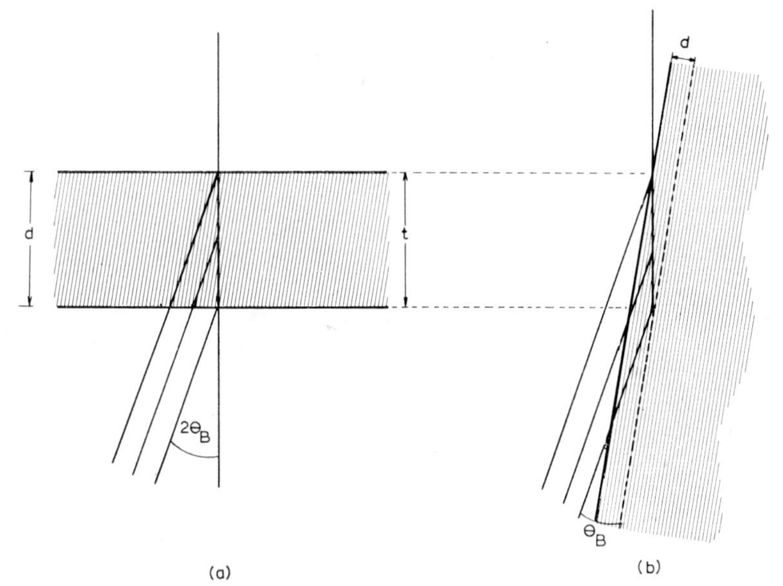

Fig. 6.3. Electron diffraction from a crystalline specimen (*a*) in transmission and (*b*) in reflection. The effective thickness of material from which the diffraction information is collected, *d*, is much reduced by using the reflection technique.

The second method of studying the arrangement of atoms at the surface of a solid is low energy electron diffraction (LEED). This technique relies on the use of very slow electrons (a few tens to a few hundreds of electron volts) to reduce penetration into the specimen. Since the conventional electron microscope cannot accurately produce a beam of electrons of such low energy and, even more important, since the imaging of the diffraction pattern is best achieved in an entirely different way, LEED is normally carried out in special purpose-built apparatus. Since electrons of such low energy have very long wavelengths the angles through which the electrons are scattered may be quite large. Consequently the diffraction pattern is viewed on a hemispherical screen, as shown in fig. 6.4. The patterns arise from only the first few atomic layers of the specimen surface and can be strongly affected by the presence of adsorbed gases. This has meant that LEED has become a powerful technique for the study of adsorption but also that very high vacuum techniques are necessary to produce an initial clean surface and maintain controlled adsorption conditions.

From this very brief survey of a few of the many possible techniques for investigating the arrangements of atoms in solids it should at least

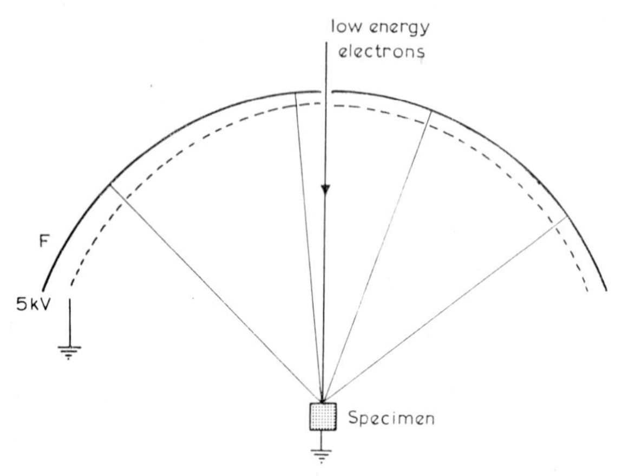

Fig. 6.4. The experimental set-up for low energy electron diffraction (LEED).

be clear that the methods which can be carried out in electron microscopes are extremely valuable, particularly since the techniques are mostly very simple. However, they are never going to supplant the essentially complementary X-ray and LEED methods, which give us information on a scale which the electron microscope cannot.

6.3. *Composition determination*

In the general field of 'analysis' there has for many years been a very wide range of techniques available, from the burette and beaker costing a pound or two to chromatographs or mass spectrometers costing tens or even hundreds of thousands of pounds. Until recently the electron probe microanalyser fitted into this broad spectrum as an expensive instrument which fulfilled a unique role, since it permitted the non-destructive analysis of very small volumes which could be precisely located and often 'seen'. More recently the uniqueness of the method has been challenged by the development of a large number of techniques which offer some advantage over EPMA in particular areas. We could perhaps summarize the capabilities of 'standard' EPMA as being the not very accurate ($\pm 1\%$) analysis and imaging of fairly small volumes (1×10^{-18} m^3 in a solid specimen; perhaps 1×10^{-23} m^3 in a thin specimen) containing moderate concentrations ($> 0{\cdot}1\%$) of heavyish elements ($Z > 10$, or $Z > 4$ with more difficulty). We can then compare the potential of other techniques under the five headings *accuracy*, *imaging*, *volume analysed*, *minimum detection limit*, and *element range*. As we shall see, it is quite easy to find instruments which can better the EPMA in one or two of these areas, but it is not easy to suggest a practicable technique which has an advantage in all five.

The most *accurate* analytical techniques are generally destructive and require a large sample of the material. Thus wet chemical analysis is capable of accuracy of orders of magnitude better than EPMA but cannot be considered a serious competitor because of its shortcomings in terms of imaging and the volume analysed. Similarly emission and absorption spectroscopy are simpler, quicker and cheaper than EPMA and are routinely available. However, since they involve the vaporizing of the portion of the specimen to be analysed (by a spark or a laser beam) they are destructive and there is no possibility of imaging the analysed area. Mass spectrometry too suffers from these disadvantages, although in terms of sensitivity and element range it is hard to match. There are, however, a number of physical techniques which are non-destructive, which restrict the volume analysed and may also offer imaging as well. Since each of these techniques offers some particular advantage over EPMA it is worth considering them in slightly more detail and also considering to what extent electron probe techniques might develop to match the newcomers.

First let us consider the range of instruments which use electrons or X-rays to excite the atoms of the specimen and then analyse the subsequently emitted radiation to determine the composition of the analysed volume. Figure 6.5 illustrates, in terms of an atom of the specimen, how four of these methods work. Unfortunately at this stage we also need to introduce still more sets of initials, as if TEM, SEM, EPMA, EMMA, WDS and EDS were not enough!

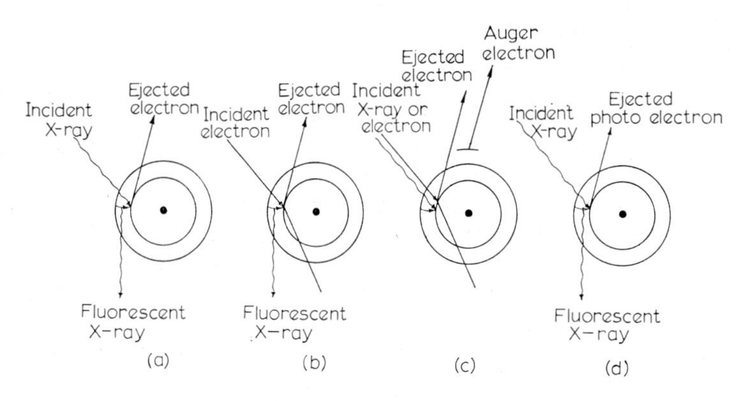

Fig. 6.5. The transitions within an atom which are used in the techniques of (*a*) X-ray fluorescence; (*b*) electron probe microanalysis; (*c*) Auger electron spectroscopy; and (*d*) X-ray photon spectroscopy (ESCA).

It is in principle possible to knock electrons out of their atomic shells, and hence give rise to the emission of characteristic X-rays, by bombarding the specimen with a variety of particles or radiations. The two most commonly used for analysis are X-ray fluorescence (XRF) illustrated in fig. 6.5 (*a*) and electron probe microanalysis (EPMA) illustrated in fig. 6.5 (*b*). XRF is a very widely used technique and in terms of the analysis of the emitted X-radiation is very similar to EPMA. However, since the beam of X-rays used to excite fluorescence cannot be focused or scanned, and since it penetrates a relatively large distance into the specimen, the XRF technique cannot be used to analyse small imaged volumes and it is usually reserved for a quick routine analysis of fairly bulky specimens.

The major alternative to the study of characteristic X-ray emission from a specimen is to analyse the energy of the secondary electrons emitted. Two methods which are currently extremely popular are illustrated in fig. 6.5 (*c*) and (*d*). An Auger electron can be emitted instead of an X-ray when an excited atom decays and an inner shell ' hole ' is filled by an outer electron—in fact this process occurs rather more frequently than X-ray emission. The Auger electron has an energy consisting of the difference in energy between the two electron shells in fig. 6.5 minus its own binding energy. The kinetic energy of each Auger electron is therefore characteristic of the atomic species in the specimen which emitted it. Auger electrons can be excited by a primary beam of either X-rays or electrons, which means that it should be possible to analyse specimens in an SEM. This has been done, although it is difficult because of the low energy of the Auger electrons—they do not travel far unless the vacuum is very good (about 10^{-9} Torr or better) and they are easily deflected by stray magnetic fields.

The final technique in this group is X-ray photoelectron spectroscopy (XPS) commonly known as ESCA (Electron Spectroscopy for Chemical Analysis). This method employs a monochromatic (single wavelength) beam of X-rays (normally Al K_α or Mg K_α) to excite inner shell electrons. The kinetic energy of these photoelectrons is then given by the energy of the X-ray photons, E_x, minus the electron binding energy and is also, therefore, characteristic of the atoms emitting them. There are two great features about AES (Auger Electron Spectroscopy) and XPS: the electron energies can be measured so accurately that differences in electron binding energy can be detected from the same chemical element when it is in different states of combination. For example, the photoelectrons from aluminium metal will have a detectable difference in energy from those from alumina (Al_2O_3). It is therefore possible to analyse a specimen not only for its constituent atoms but also for their chemical state, which is extremely useful. The second great advantage (or in some cases disadvantage) of these techniques is that the electrons being analysed come only from the top few atomic layers of the specimen and therefore the slightest covering of grease, oxide or whatever will be detected. Clearly it is not possible to image the specimen in an XPS instrument since the primary X-ray beam cannot be focused or scanned. This ought to be possible for AES using an electron beam to excite the Auger electrons but at present so few electrons are emitted (i.e. the signal is so small) that Auger electron composition images similar to the X-ray images formed in EPMA take a very long time to create. However, this is an area where development is rapid and we can expect significant improvements in the near future.

There is a second range of instruments coming into use which can offer many of the advantages of EPMA. These instruments use beams of ions in place of the electron or X-ray beams we have so far considered. The advantages which are hoped for arise from two main factors: the scattering of an ion beam within a solid specimen occurs typically over a smaller volume than that which we have come to expect with electrons (e.g. fig. 5.1); the region to be analysed may therefore be shallower than in the EPMA. The second factor is that secondary ions can be generated by bombardment of a solid specimen with a primary ion beam and, therefore, since ions may be compounds or even different atomic isotopes, we should in principle be able to analyse not only for elements but also for isotopes or chemical compounds. Figure 6.6 illustrates, in a very much simplified way, the three main types of ion probe technique.

The proton beam microanalyser (fig. 6.6 (*a*)) is exactly analogous to the EPMA. A beam of hydrogen ions (protons) is used to excite characteristic X-rays in much the same way as does the electron beam of the conventional EPMA. The X-rays which are emitted can then be analysed in either a wavelength-dispersive or energy-dispersive detector. The advantages of the technique are that the depth of penetration of the protons is not as great as that for electrons while the peak-to-background ratio is very high, leading to excellent sensitivity (see Section 5.5). The proton beam can be deflected and scanned just like an electron beam and therefore the whole range of techniques described for EPMA in Chapter 5 is available. The disadvantages are however likely to limit the technique to a few special applications: the spatial resolution is worse then 5 μm and in order to generate sufficient X-rays for analysis in a reasonable time protons of several MeV energy have to be used. It is rather more difficult (and expensive) to generate and focus a beam of 3 MeV protons than one of 30 keV electrons.

The other two ion techniques use ions both to excite the specimen and as the analysing radiation. The process of erosion which occurs when an ion beam hits a solid specimen is known as *sputtering*. Atoms (or molecules) of the solid are knocked out of the surface and may be charged (i.e. ions) or uncharged (i.e. atoms or molecules). In most cases a majority of the atoms are emitted as ions and consequently can be deflected by a magnetic field. If they pass into a large uniform magnetic field (perpendicular to the paper, downwards, in fig. 6.6 (*b*). then the deflection of each ion will depend on its charge and its mass) Consequently the beam of secondary ions is split by this ' magnetic prism ' into several beams each containing ions of the same mass

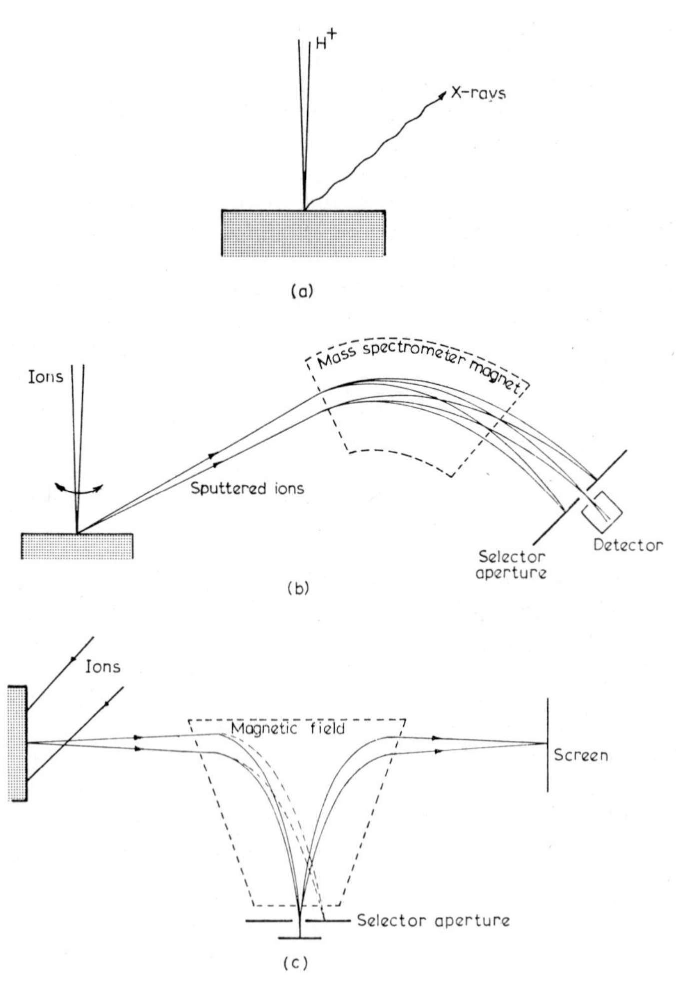

Fig. 6.6. Ion beam analytical techniques. (*a*) Proton beam X-ray micro-analyser; (*b*) scanning ion microprobe; (*c*) ion microscope.

(assuming for the moment that all ions are singly charged). This is basically the action of a mass spectrometer. In the ion microprobe, fig. 6.6 (*b*), if we place an aperture at the far side of the magnetic prism we can select whichever beam we choose and allow only a single type of ion to fall on to the detector and be counted. We therefore have a system which operates just like a scanning electron microscope except that the primary beam consists of ions instead of electrons and the secondary effect to be studied is the number of a particular ion species rather than the number of secondary electrons arriving at a detector. This instrument can therefore be used to display scanning images of the distribution of any element in the specimen surfaces, or alternatively the ion beam can be held on a single spot of the specimen and a qualitative analysis carried out. The three great advantages of the ion microprobe are: (1) that the sputtered ions come from only the top 1 nm of the specimen, giving it a depth resolution similar to XPS (ESCA) at the same time as a spatial resolution of the same order as the EPMA (1 μm); (2) the magnetic prism can analyse any element from hydrogen to uranium and also compound ions such as CO^- and $CH_3{}^-$; and (3) the distribution of an element in depth can be studied by continuing the sputtering for a long time—gradually the specimen is eroded and the analysis represents a deeper region of the specimen. Against these obvious advantages must be set the difficulty of obtaining a quantitative analysis and a certain number of practical difficulties. The possibility of quantitative analysis seems far off, since the sputtering rate of each atom depends critically on its chemical state. Thus the rate of copper ion sputtering from metallic copper is very different from the rate of copper ion sputtering from cuprous oxide. This is a problem which is not really encountered in EPMA and makes the ion microprobe useless at the moment for most quantitative analyses.

The third main type of ion probe technique is the ion microscope illustrated schematically in fig. 6.6 (*c*). This bears the same relationship to the ion microprobe described above as the TEM does to the SEM. The ion distribution image is formed from a whole area of specimen at once using ' lenses ' just as in any optical microscope. The specimen is bombarded with a broad beam of primary ions and the sputtered ions enter a mass spectrometer exactly as in the ion microprobe; this enables any chosen ion to be selected by an aperture. Here any similarity to the ion microprobe ends. After passing through the selector aperture the chosen ions are reflected back through the magnetic prism by an electrostatic field which acts as a mirror. The ion beam is now deflected in the opposite direction, away from the specimen; it passes through a projector lens and eventually forms an image on a fluorescent screen. The magnetic prism and electrostatic mirror combine to act exactly like a lens and hence every point in the specimen corresponds to a point in the image. The instrument is therefore an optical microscope which can use ions of any mass to form an image with resolution better than 1 μm which corresponds to the distribution of that atomic species in the specimen. Most of the advantages and disadvantages already discussed in relation to the ion microprobe also apply to the ion microscope. Nonetheless its great advantages remain that it can operate using even the lightest elements, for instance hydrogen or lithium, which are inaccessible to EPMA, and that it is extremely sensitive to small concentrations of elements near or on the specimen surface. It has been claimed that in favourable circumstances as little as one part per million of a foreign atom in a single atomic layer can be detected on a surface.

Of the techniques which can be used to analyse regions of a specimen as small as 10 μm in diameter none has yet been developed to such an advanced level as EPMA and only one (the proton beam microanalyser) is yet capable of a quantitative accuracy anywhere near as good as ±1%. The combination of performance available with an EPMA therefore seems unchallenged, but several other techniques offer the possible advantage of shallow depth of analysis and the undoubted advantages of increased element range and the possibility of distinguishing between identical atoms in different chemical environments. To what extent is the electron probe microanalyser capable of development in these directions? In brief the answer is probably ‘ not very far ’: the very nature of the electrons used for excitation and the X-rays used for analysis preclude the analysis to a layer of the specimen very near the surface, since both types of radiation penetrate the specimen quite deeply. Differentiation between the various possible chemical environments of a single atomic species does not seem likely to be achieved using EPMA either. Although the chemical bonds in which an atom is involved do alter the energy and hence wavelength of the X-rays it emits, these effects are only large enough to detect in the electron probe microanalyser for very light elements, e.g., carbon and oxygen. It is almost always going to be more efficient to carry out this sort of analysis using an XPS instrument or, if good spatial resolution is essential, in an SEM with facilities for Auger electron analysis.

However, it seems that for a few years at least EPMA (of solid or thin specimens) will be the best way to resolve concentration differences between very small regions of a specimen, for instance, between a precipitate and its matrix, or between a cell and its nucleus. It will be some time before any comparable technique can analyse small volumes to an accuracy of about ±1%.

As an example of just how far it is possible to develop a technique if you try, we shall close this chapter by describing a method of determining the nature of a single atom. It must be emphasized that this is far from being a routine operation and has only been performed in a very few highly specialized laboratories in the world. The technique is commonly known as the ' atom probe '. You will recall that the field ion microscope (fig. 6.1) enables an image to be formed in which each atom in the tip of a very fine specimen is seen as a bright dot on a large hemispherical screen. The principle of the atom probe is to allow one of these bright spots to fall on a small hole, say at H in fig. 6.1 (*a*), which leads to a very sophisticated mass spectrometer. The voltage on the specimen is now pulsed so that approximately one layer of atoms is ejected from the surface of the specimen tip. The newly formed ions will travel in straight lines to the fluorescent screen. The single chosen atom will pass through the hole H and can be analysed in a special type of mass spectrometer. The nature of a single pre-selected atom can therefore be determined. This is clearly an extraordinary technique, which is extremely difficult to use, but it does give an indication of the ultimate in analysis.

SUGGESTIONS FOR FURTHER READING

Optics and light microscopy:

M. Born and E. Wolf, *Principles of optics*, 4th edition, Pergamon Press, 1970.
S. Bradbury, *The evolution of the microscope*, Pergamon Press, 1967.
R. W. Ditchburn, *Light*, Blackie, 1963.
L. C. Martin, *The theory of the microscope*, Blackie, 1966.

General texts on electron beam techniques:

J. A. Belk and A. L. Davies, *Electron microscopy and microanalysis of metals*, Elsevier, 1968.
J. K. Koehler, *Advanced techniques in biological electron microscopy*, Springer-Verlag, 1973.
L. E. Murr, *Electron optical applications in materials science*, McGraw-Hill, 1970.
V. A. Phillips, *Modern metallographic techniques*, John Wiley, 1971.

Transmission electron microscopy:

A. M. Glauert (ed.), *Practical methods in electron microscopy*, series of volumes, North-Holland, 1972, 1974, etc.
P. J. Goodhew, *Specimen preparation in materials science*, North-Holland, 1972.
G. H. Haggis, *The electron microscope in molecular biology*, Longmans, 1968.
M. A. Hayat, *Principles and techniques of electron microscopy: biological applications*, Vol. 1, 1970, Vol. 2, 1972, Van Nostrand Reinhold.
P. B. Hirsch, A. Howie, R. B. Nicholson, D. W. Pashley and M. J. Whelan, *Electron microscopy of thin crystals*, Butterworths, 1965.
B. E. Juniper, A. J. Gilchrist, G. C. Cox and P. R. Williams, *Techniques for plant electron microscopy*, Oxford, Blackwell, 1970.
G. A. Meek, *Practical electron microscopy for biologists*, Wiley-Interscience, 1970.
E. H. Mercer and M. S. C. Birbeck, *Electron microscopy; a handbook for biologists*, 3rd edition, Blackwell, 1972.
D. F. Parsons, *Some biological techniques in electron microscopy*, Academic Press, 1970.
F. Scanga, *Atlas of electron microscopy: biological applications*, Elsevier, 1964.
P. R. Swann *et al.* (ed.), *High voltage electron microscopy*, Academic Press, 1974.
B. S. Weakley, *A beginner's handbook in biological electron microscopy*, Churchill Livingstone, 1972.

Scanning electron microscopy:

J. W. S. Hearle, J. T. Sparrow and P. M. Cross, *The use of the scanning electron microscope*, Pergamon Press, 1972.
V. H. Heywood (ed.), *Scanning electron microscopy: systematic and evolutionary applications*, Academic Press, 1971.
C. W. Oatley, *The scanning electron microscope*, Cambridge University Press, 1972.

Electron probe microanalysis:

C. A. Anderson (ed.), *Microprobe analysis*, Wiley, 1973.
D. R. Beaman and J. Isasi, *Electron beam microanalysis*, A.S.T.M. Special technical publication 506, 1972.
T. D. McKinley, K. F. J. Heinrich and D. B. Wittry, *The electron microprobe*, Wiley, 1966.
J. C. Russ (ed.), *Energy dispersive X-ray analysis*, A.S.T.M. Special technical publication 485, 1971.

Other areas:

B. E. P. Beeston, R. W. Horne and R. Markham, *Electron diffraction and optical diffraction techniques*, North-Holland, 1972.
K. M. Bowkett and D. A. Smith, *Field ion microscopy*, North-Holland, 1970.
E. W. Müller and T. T. Tsong, *Field ion microscopy: principles and applications*, Elsevier, New York, 1969.

INDEX

THE WYKEHAM SCIENCE SERIES

1 †*Elementary Science of Metals* — J. W. Martin and R. A. Hull
2 *Neutron Physics* — G. E. Bacon and G. R. Noakes
3 †*Essentials of Meteorology* — D. H. McIntosh, A. S. Thom and V. T. Saunders
4 *Nuclear Fusion* — H. R. Hulme and A. McB. Collieu
5 *Water Waves* — N. F. Barber and G. Ghey
6 *Gravity and the Earth* — A. H. Cook and V. T. Saunders
7 *Relativity and High Energy Physics* — W. G. V. Rosser and R. K. McCulloch
8 *The Method of Science* — R. Harré and D. G. F. Eastwood
9 †*Introduction to Polymer Science* — L. R. G. Treloar and W. F. Archenhold
10 †*The Stars; their structure and evolution* — R. J. Tayler and A. S. Everest
11 *Superconductivity* — A. W. B. Taylor and G. R. Noakes
12 *Neutrinos* — G. M. Lewis and G. A. Wheatley
13 *Crystals and X-rays* — H. S. Lipson and R. M. Lee
14 †*Biological Effects of Radiation* — J. E. Coggle and G. R. Noakes
15 *Units and Standards for Electromagnetism* — P. Vigoureux and R. A. R. Tricker
16 *The Inert Gases: Model Systems for Science* — B. L. Smith and J. P. Webb
17 *Thin Films* — K. D. Leaver, B. N. Chapman and H. T. Richards
18 *Elementary Experiments with Lasers* — G. Wright and G. Foxcroft
19 †*Production, Pollution, Protection* — W. B. Yapp and M. I. Smith
20 *Solid State Electronic Devices* — D. V. Morgan, M. J. Howes and J. Sutcliffe
21 *Strong Materials* — J. W. Martin and R. A. Hull
22 †*Elementary Quantum Mechanics* — Sir Nevill Mott and M. Berry
23 *The Origin of the Chemical Elements* — R. J. Tayler and A. S. Everest
24 *The Physical Properties of Glass* — D. G. Holloway and D. A. Tawney
25 *Amphibians* — J. F. D. Frazer and O. H. Frazer
26 *The Senses of Animals* — E. T. Burtt and A. Pringle
27 †*Temperature Regulation* — S. A. Richards and P. S. Fielden
28 †*Chemical Engineering in Practice* — G. Nonhebel and M. Berry
29 †*An Introduction to Electrochemical Science* — J. O'M. Bockris, N. Bonciocat, F. Gutmann and M. Berry
30 *Vertebrate Hard Tissues* — L. B. Halstead and R. Hill
31 †*The Astronomical Telescope* — B. V. Barlow and A. S. Everest
32 *Computers in Biology* — J. A. Nelder and R. D. Kime
33 *Electron Microscopy and Analysis* — P. J. Goodhew and L. E. Cartwright
34 *Introduction to Modern Microscopy* — H. N. Southworth and R. A. Hull
35 *Real Solids and Radiation* — A. E. Hughes, D. Pooley and B. Woolnough
36 *The Aerospace Environment* — T. Beer and M. D. Kucherawy
37 *The Liquid Phase* — D. H. Trevena and R. J. Cooke
38 *From Single Cells to Plants* — E. Thomas, M. R. Davey and J. I. Williams
39 *The Control of Technology* — D. Elliot and R. Elliot

THE WYKEHAM TECHNOLOGY SERIES

1 *Frequency Conversion* — J. Thomson, W. E. Turk and M. J. Beesley
2 *Electrical Measuring Instruments* — E. Handscombe
3 *Industrial Radiology Techniques* — R. Halmshaw
4 *Understanding and Measuring Vibrations* — R. H. Wallace
5 *Introduction to Tribology* — J. Halling and W. E. W. Smith

All orders and requests for inspection copies should be sent to the appropriate agents. A list of agents and their territories is given on the verso of the title page of this book.

† (*Paper and Cloth Editions available.*)